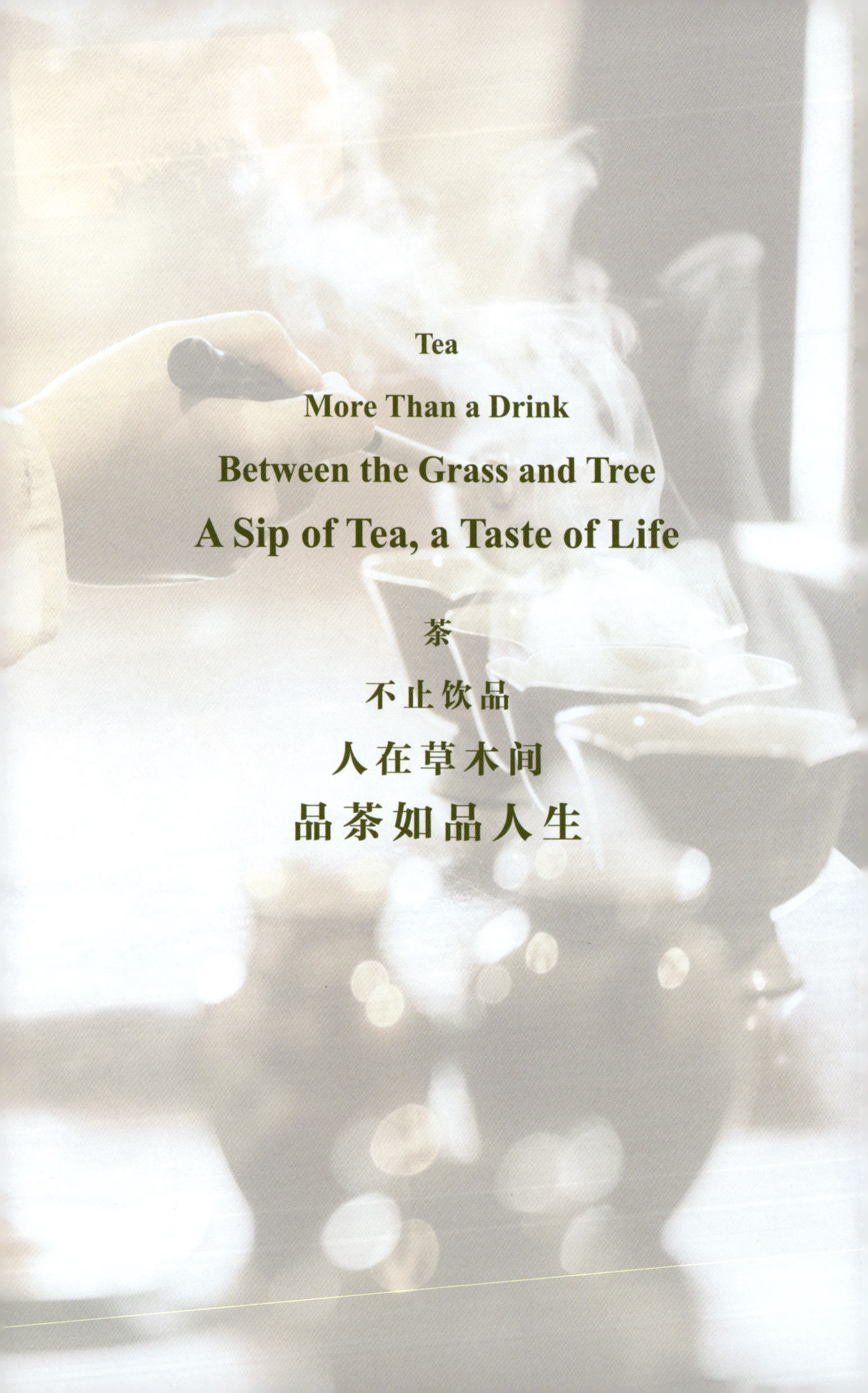

Tea

More Than a Drink

Between the Grass and Tree

A Sip of Tea, a Taste of Life

茶

不止饮品

人在草木间

品茶如品人生

用英语讲好中国非遗系列

茶事如斯

TEA MATTERS

朱宁 编著

北京·旅游教育出版社

图书在版编目（CIP）数据

茶事如斯 = Tea Matters：汉英对照 / 朱宁编著.
北京：旅游教育出版社，2025. 4. --（用英语讲好中国
非遗系列）. -- ISBN 978-7-5637-4866-2

Ⅰ. TS971.21

中国国家版本馆CIP数据核字第2025L4F533号

用英语讲好中国非遗系列

茶事如斯
Tea Matters

朱宁　编著

策　　划	李红丽
责任编辑	李红丽
出版单位	旅游教育出版社
地　　址	北京市朝阳区定福庄南里 1 号
邮　　编	100024
发行电话	（010）65778403　65728372　65767462（传真）
本社网址	www.tepcb.com
E - mail	tepfx@163.com
排版单位	北京旅教文化传播有限公司
印刷单位	天津雅泽印刷有限公司
经销单位	新华书店
开　　本	880毫米×1230毫米　1/32
印　　张	10.75
字　　数	218 千字
版　　次	2025 年 4 月第 1 版
印　　次	2025 年 4 月第 1 次印刷
定　　价	58.00 元

（图书如有装订差错请与发行部联系）

前　言
Foreword

2022 年 11 月 29 日，"中国传统制茶技艺及其相关习俗"被列入联合国教科文组织人类非物质文化遗产代表作名录。历经数千年发展，中国形成了绿茶、红茶等六大茶类及花茶等再加工茶，共 2000 余种茶品，其制作技艺与饮茶习俗融入百姓生活，贯穿婚嫁祭祀、节令庆典。以丝绸之路、茶马古道等古代商路为纽带，茶及其文化穿越历史、跨越国界，成为中国与世界人民相知相交、中华文明与世界其他文明交流互鉴的重要媒介，成为人类文明共同的财富。

茶，承载着中华数千年的文化；茶，飘香于世界的各个角落。茶是中国的，也是世界的，茶是讲好中国故事最好的载体之一。《茶事如斯》一书以茶为主题，以中英双语为表达形式，旨在通过茶的历史文化、传播发展、守正创新等内容，呈现中华文明的博大精深和历久弥新，彰显各美其美、美美与共的文化自信。

本书立足于茶文化的"国际化""年轻化"，分为四大部分十个小项内容。四大部分包括：茶出中国、茶韵风雅、茶养生气、茶香四海；十个小项包括：茶之源、茶之类、茶之泡、茶之具、茶之饮、茶之藏、茶之效、茶之馔、茶之传、茶之俗。在设计思路上，全书以跨文化传播为目标，前三部分勾勒了茶的历史源流、文化雅趣与健康智慧，最后一部分则详述了茶的全球迁徙与多元文化呈现。笔者希望通过这样的架构，能与读者共同探寻在全球化背景下中国茶文化的传统魅力和当代价值。

在写作过程中，笔者查阅了大量国内外关于茶文化、茶健康、茶产业发展的资料，参考了许多品牌的官网资料，并引用了一些专家学者的成果和观点，在此对他们深表谢意。

本书得以出版离不开上海科学技术出版社周星娣老师、旅游教育出版社李红丽老师的鼓励与支持；也离不开赵德芳、慎丹丹、高锋、吴江华、马娜、云晴、陈燕、姚建静、黄佳蕴、龚伟、吴忱等的帮助，他们知我所需，解我所难，提供了多张精美图片。此外，年轻的茶人刘栅杉、朱子彤协助了部分资料的收集和图片的拍摄，在此一并表示感谢。

由于笔者水平有限，书中难免存在疏漏和不足之处，恳请读者批评指正。

朱 宁

2025 年 3 月

目　录
Contents

茶事如斯 TEA MATTERS

第四部分　茶香四海
Connecting the World with Tea

引子：茶之性
Introduction: The Nature of Tea

在长达数千年的漫长岁月里，茶以各种形态出现，它曾是药物、食物、饮品，被当作每日的必需品、文化消费品、祭祀品、皇室贡品、特权阶级专属饮品等。它具有品饮性、艺术性、健康性、社交性等多重属性，而其特有的经济属性，使其以"蝴蝶效应"之势，影响了多个国家和地区的政治、经济和社会变迁。或许没有任何其他饮品或食物有如此多样的功效，能够如此深远地改变世界的历史进程。

For thousands of years, tea has played an important role in human civilization, serving as medicine, food, and beverage. It has been a daily essential, a cultural symbol, a sacrificial offering, a royal tribute, and an exclusive indulgence for the elite. Tea possesses multiple attributes, including potability, artistry, health benefits, and social significance. Its economic potency once triggered butterfly effects, influencing political, economic, and social transformations across multiple countries and regions. No other beverage or food has wielded such a profound and far-reaching influence.

柴米油盐酱醋茶——品饮性
Tea as a Daily Essential: The Drinkability Factor

茶不仅融入中国人的日常生活，也是全球数十亿人生活中不可或缺的部分。当下，全球已有 160 多个国家和地区有饮茶习惯，饮茶人口超过 30 亿人，每天茶叶消耗量高达 30 亿杯，茶已成为仅次于水的全球第二大饮品。

As we know, tea is one of the seven essentials of daily life for Chinese, along with firewood, rice, oil, salt, sauce and vinegar. However, tea is not just an integral part for Chinese people—it is a global phenomenon. Today, tea is enjoyed in over 160 countries and regions, with a global tea-drinking population exceeding 3 billion people. The world consumes an astonishing 3 billion cups of tea per day, making it the second most consumed beverage after water.

无论茶承载多少光环，本质上，茶是一种饮品，其品饮性既是这片神奇的树叶最原始的功能，也是其最朴素的存在意义。

Despite its deep cultural and historical significance, tea's most fundamental function remains its drinkability—a simple yet timeless pleasure that has endured for millennia.

琴棋书画诗酒茶——艺术性
Tea as an Art: A Cultural and Aesthetic Expression

在中国，自古以来，茶不仅是日常饮品，更是与琴、棋、书、画、诗、酒一起，成为人们的精神食粮。

In China, tea is more than just a beverage—it is a cultural and spiritual pursuit. Alongside music, literature, and fine arts, tea has long been an essential part of the Chinese way of life.

早在西晋时期，杜育的《荈赋》便首次以诗赋形式赞颂茶叶；唐代，陆羽提出的"为饮，最宜精行俭德之人"首次赋予茶人文内涵；宋代，范仲淹的诗句"胜若登仙不可攀，输同降将无穷耻"，描绘了彼时斗茶雅集之趣；明代，张源的《茶录》提出的"饮茶以客少为贵。众则喧，喧则雅趣乏矣"等品饮规范，更将饮茶上升为一种身心体验。

As early as the Western Jin Dynasty, Du Yu's poem *Chuanfu (Ode to Tea)* was among the first literary works to praise tea. During the Tang Dynasty, Lu Yu's *The Classic on Tea* established tea as a symbol of simplicity, virtue, and self-discipline. By the Song Dynasty, tea gatherings had become a refined social ritual, as reflected in Fan Zhongyan's poetry. In the Ming Dynasty, Zhang Yuan emphasized in his work *The Record of Tea* that tea drinking is best enjoyed in a quiet and intimate setting, stating: *"it is important in drinking tea that the guests be few. Many guests would make it noisy, and noisiness takes away from its cultured charm."*[①]

的确，对中国人而言，饮茶是物质享受，更是精神享受，是一种仪式，一种境界，一种精致品位的展示。在诗情画意的氛围中，或独处小憩，或与三五好友谈笑，怡情悦性，不亦乐乎。

For the Chinese people, tea drinking is not merely a physical indulgence—it is a spiritual practice, a ceremonial experience, and an expression of refined taste. Whether enjoyed alone for meditation and self-reflection or shared with close friends in a poetic atmosphere, tea embodies a harmonious blend of culture and tranquility.

茶为万病之药——健康性
Tea as Medicine: A Path to Wellness

对许多西方人而言，茶是提神的饮品；但对中国人而言，茶是万

---

① 此段文字引自林语堂的哲学著作 *The Importance of Living*，原著用英文写成，于1937年在美国首次出版，中文版本译为《生活的艺术》。

灵之药。人类对茶的最初认知与其健康属性紧密关联。数千年来，茶是最贴近中国人生活的药食两用植物。但在 19 世纪前，其科学原理一直扑朔迷离。

To many Westerners, tea is simply a refreshing beverage, but for the Chinese, tea is a panacea. Humanity's earliest understanding of tea was closely tied to its medicinal properties. For thousands of years, tea has been a dual-purpose plant, both a dietary staple and a natural remedy. However, before the 19th century, its scientific properties remained largely a mystery.

20 世纪后半叶，茶与健康的研究起步，茶学迎来了自然科学理论的探秘时代。茶的主要成分如儿茶素、茶黄素、茶氨酸、咖啡碱等被逐一确认，至今已经鉴定的物质达到 1500 多种。随着化学工程、分子生物学、细胞生物学、生物化学、食品营养学和临床医药等大量先进科学技术参与茶的研究，茶与健康得到空前关注。①

It wasn't until the latter half of the 20th century that researchers began exploring the scientific principles behind tea's health benefits. Studies have identified key components such as catechins, theaflavins, theanine, and caffeine, among over 1,500 bioactive compounds. With advancements in chemical engineering, molecular biology, cell biology, food nutrition, and clinical medicine, tea's health benefits have gained unprecedented recognition.

当下，人们对于健康生活的要求越来越迫切。虽然茶的健康属性毋庸置疑，但在一定程度上，其健康属性的发挥也要求人们对茶进行科学而合理的利用。

Today, as people prioritize healthy lifestyles, the demand for natural and

---

① 关于茶学发展的内容，参考屠幼英教授在《茶与健康》一书中的表述，略有缩减。

functional beverages has surged. While tea's health benefits are well-established, its effectiveness also depends on scientific and mindful consumption.

举杯共叙，茶香融情——社交性 [①]
Tea as a Social Catalyst: Its Social Essence

茶自古就有浓厚的社交属性。茶不仅是日常生活中的待客之道，也是人们交流情感、增进友谊的重要媒介。在中国，茶馆自古就是社交场所，人们在此喝茶聊天，既可打发时间，也可商务洽谈。在西方，19世纪兴起的"英式下午茶"文化更是将茶的社交属性推向极致，在幽雅的环境中品茶聊天，成为一种独特的社交仪式。

Tea has long been deeply embedded in social culture. Beyond being a simple gesture of hospitality, it serves as a bridge for connection and friendship. In China, tea houses have traditionally been vibrant social hubs where people gather over tea—whether to pass the time or engage in business discussions. In the West, the rise of English afternoon tea in the 19th century elevated tea's social role, transforming it into a refined ritual where people bonded over tea in an elegant setting.

当下，新式茶饮的流行进一步强化了茶的社交属性。各类茶饮店成为年轻人聚会、打卡的热门场所。茶，不仅是一杯饮品，更是一种

[①] 林语堂先生在《生活的艺术》（越裔，译，2017）一书中对此是这么说的，"我以为从人类文化和快乐的观点论起来，人类历史中的杰出新发明，其能直接有力地有助于我们的享受空闲、友谊、社交和谈天者，莫过于吸烟、饮酒、饮茶的发明。这三件事有几样共同的特质：第一，它们有助于我们的社交；第二，这几件东西不至于一吃就饱，可以在吃饭的中间随时吸饮；第三，都是可以借嗅觉去享受的东西。它们对于文化的影响极大，所以餐车之外另有吸烟车，饭店之外另有酒店和茶馆，至少在中国和英国，饮茶已经成为社交上一种不可少的制度。"

连接人与人之间情感的纽带。

Today, the rise of new-style tea drinks has further strengthened tea's social appeal. Trendy tea shops have become go-to spots for young people to meet. Tea is no longer just a drink; it has become a cultural symbol that brings people together.

当然，在任何时候，茶社交属性的实现都是需要条件的，正如林语堂先生所认为的，茶的社交属性的实现需要合适的友伴和环境，只有在相宜的情调下，与趣味相投的人共饮，茶才能促进友谊的加深，才能真正成为一种享受。

However, tea's social function is not automatic—it flourishes under the right conditions. As the renowned Chinese writer Lin Yutang observed, the enjoyment of tea is best experienced in the company of kindred spirits and within an inviting atmosphere. Only when shared with like-minded people in the right setting can tea truly foster deeper connections and become a source of genuine pleasure.

不可或缺，绿色黄金——经济性
Tea as Liquid Gold: Its Economic Significance[①]

在世界的文明进程中，茶的属性也发生了变化。随着文化和贸易的日趋频繁，茶的经济性（或称货币属性）逐渐凸显。恰如《绿色黄金：茶叶帝国》一书中所写："茶叶货币制度非常完美、贴切地表现出了金钱的核心功能。即计算价值的单位、交换媒介与财富积累和储蓄等。它很轻，可以做成规格化的茶砖，而且很有价值。茶还有一点优

---

① 茶的经济性英文部分主要参考自戎新宇的 *Tea Nation*（2018），中文版《茶的国度：改变世界进程的中国茶》于 2019 年在国内出版。

于银币和纸钞，那就是，如果有需要，还可以拿来食用和饮用。"

Tea's economic role has transformed significantly over time. As cultural exchanges and trade flourished, tea became increasingly recognized for its monetary and commercial value. In *Green Gold: The Tea Empire*, the author stated: "The tea currency system perfectly and relevantly manifested the core functions of money as a value calculation unit, an exchange medium as well as a way of wealth accumulation and saving. It is light and can be made into standard tea disks. It is very valuable. Tea has another merit that silver coins and paper money do not have: it can be eaten and drunk, if needed."

戎新宇写道，"起源于中国的茶叶，就如同苏丹的咖啡、土耳其的葡萄酒一样，是人类文明史上所创造出的宝贵财富中为数不多且可被饮用的'液体黄金'。"

As a Chinese scholar named Rong Xinyu wrote, *"Just like Sudanese coffee and Turkish wine, Chinese tea is one of humanity's greatest treasures— a drinkable form of liquid gold."*

茶和天下，美美与共——外交性
Tea and Diplomacy: A Cultural Ambassador

茶，这片神奇的东方树叶，其外交属性很早就被发掘和利用。我们姑且将茶外交分为民间外交和官方外交两部分。

Tea has long served as a bridge for international relations. Historically, tea diplomacy can be divided into two categories: people-to-people exchanges and official state diplomacy.

茶民间外交属性主要体现在贸易、文化交流领域。茶叶是古代丝绸之路三大贸易商品之一；早在唐代，茶就通过遣唐使、僧人传播到东亚各国，在随后的宋元明清，中华茶文化不断影响着周边国家；17世纪初，荷兰辗转将茶叶运至欧洲，开启了欧洲的饮茶之风。

Tea has long played a vital role in grassroots diplomacy, particularly in the areas of trade and cultural exchange. As one of the three key commodities along the ancient Silk Road, tea served as a cultural bridge. As early as the Tang dynasty, it spread to East Asian countries through Japanese envoys and Buddhist monks. Throughout the following dynasties, Chinese tea culture continued to shape and inspire neighboring societies. In the early 17th century, Dutch traders brought Chinese tea to Europe via maritime routes, sparking a tea-drinking craze that soon swept across the continent.

茶的官方外交属性形成时间也较为久远。始于唐代的"茶马互市"促进了中原地区和西北边疆游牧民族的友好往来，推动了中华民族多民族共同体的形成，而与"茶马互市"密切相关的"茶马古道"不仅是经贸之道、文化之道，更是战略之道；明代郑和下西洋也推动了茶文化的进一步传播；明代末期，茶作为国礼被赠送给俄国沙皇。

In terms of official diplomacy, the Tea-Horse Trade, which began in the Tang Dynasty, facilitated friendly exchanges between the Central Plains region and the nomadic tribes of the northwestern frontier regions, fostering the formation of a multi-ethnic community within the Chinese nation. The Tea Horse Road became not just a trade route, but also a strategic and cultural pathway. In the Ming Dynasty, Zheng He's maritime expeditions further facilitated the spread of tea, and in the late Ming Dynasty tea was presented as an official diplomatic gift to foreign rulers.

中华人民共和国成立后，历代领导人都十分重视茶外交。毛泽东主席、周恩来总理多次以中国名茶为国礼，赠送来访贵宾 ①。

Modern Chinese leaders have continued this tradition of tea diplomacy. Chairman Mao Zedong and Premier Zhou Enlai frequently presented China's finest teas as national gifts to visiting dignitaries.

进入中国特色社会主义新时代以来，习近平总书记在多个重大外交场合，以茶会友、以茶论事。茶作为传递友好信息的载体亮相国家外交舞台，既彰显了中国的文化自信，也向世界传达了在多元文化下"和而不同"的理念。

In the new era, President Xi Jinping has embraced tea diplomacy on numerous major diplomatic occasions, using tea to foster friendships and facilitate discussions. As a symbol of goodwill, tea has taken center stage in China's diplomatic engagements, reflecting the nation's cultural confidence and conveying to the world the philosophy of "harmony in diversity" within a multicultural landscape.

以茶为约，信守承诺——契约性
Tea as a Covenant: A Symbol of Commitment and Trust

茶在中国传统的婚姻习俗中，发挥着契约性的作用，被视为信任、尊重与盟约的象征。如在订婚时的"下茶"仪式中，女方收下定亲茶即表示认可婚事，此时茶是婚约确立的象征；在迎亲中的"定茶"（也称奉茶）礼中，茶是家庭和谐的期盼；在同房时的"合茶"仪式上，

---

① 中华人民共和国成立之初，毛泽东主席出访苏联，就以景德镇茶具、祁门红茶、西湖龙井茶等为国礼；20世纪60年代，茶叶帮助开启了中非建交的大门；1972年，美国总统尼克松访华，毛主席以大红袍为礼；周恩来总理也曾多次以西湖龙井茶为国礼赠送来访贵宾。

茶是"与子偕老"的承诺。

Tea holds a symbolic and contractual role in traditional Chinese marriage customs, representing trust, respect, and commitment. In the engagement, the acceptance of tea signifies the bride's family's formal acknowledgment of the marriage, marking the establishment of the engagement. During the wedding, serving tea to in-laws expresses a wish for family harmony and signifies the bride's official acceptance into the groom's family. The following tea-sharing ceremony represents a solemn vow of loyalty and lifelong devotion between husband and wife.

茶树四季常青、不凋不谢,"茶不移本,植必子生",茶的这些特点与人们对婚姻从一而终、幸福美满的期盼不谋而合。茶承载着婚姻的承诺、家庭的认可及夫妻间的忠贞,体现了中华文化中重承诺、守信用的价值观。

Tea plants remain evergreen, and it is believed that they could only spout from seeds; and once transplanted, they would die. These characteristics align with the ideals of steadfast love, family continuity, and unwavering commitment in marriage. Through these rituals, tea carries the promise of union, familial recognition, and marital fidelity, reflecting the Chinese cultural values of commitment, integrity, and trustworthiness.

茶禅一味,悟道清心——宗教性
Tea and Religion: A Spiritual Connection

茶承载着深厚宗教意蕴。茶的宗教性首先体现在其与佛教的紧密联系上。千百年来,佛教视茶为修禅悟道之媒介。禅宗认为,茶能清心寡欲、祛除烦躁,使人在静心的过程中达到精神上的觉悟。因此,茶与禅宗思想紧密结合，茶不仅是饮品，更是修身养性、禅修心境的工具。

Tea is steeped in deep spiritual and religious meaning. The religious dimension of tea is most evident in its deep-rooted connection with Buddhism. For centuries, Buddhist practitioners have embraced tea as a medium for meditation and spiritual cultivation. In Zen Buddhism especially, tea is valued for its ability to clear the mind, moderate desires, and calm inner restlessness, helping practitioners achieve clarity and enlightenment through mindful stillness. As a result, tea became inseparable from Zen philosophy—elevating tea beyond a mere beverage to a powerful tool for self-discipline, inner reflection, and spiritual awakening.

此外，茶还与儒、道思想相融合。儒家通过品茶活动陶冶心智、修炼品性，茶道中的礼仪也体现了儒家的"礼治"思想。在道家思想中，茶能调和身心，具有清心、养气、提神的功效，是修炼之道的重要辅助。

In addition, tea is also closely intertwined with Confucianism and Taoism. In Confucianism, tea etiquette reflects the concept of "governing by ritual", promoting self-cultivation and discipline. In Taoism, tea is believed to harmonize the body and mind, serving as an aid in spiritual refinement.

以茶为底，万物可茗——时尚性
Tea: A Timeless Tradition with a Modern Twist

茶，这个古老的饮品，还与不同元素融合，焕发出前所未有的时尚活力。新式茶饮无疑是茶的时尚性的最好代言。新式茶饮是指采用优质茶叶、鲜奶、水果等多种食材，通过多样化的茶底和配料组合而成的现制茶饮料。

Tea, an ancient beverage, is undergoing a remarkable transformation, blending with modern elements to exude a fresh sense of style and vitality. New-style tea drinks have become the ultimate symbol of tea's fashionable appeal. Crafted with high-quality tea leaves, fresh milk, fruit, and other ingredients, these beverages offer an innovative take on traditional tea through diverse tea bases and creative combinations.

在口感配置上，新式茶饮通过独特的配方，与多种水果、饮品进行创新性"跨界"搭配，并通过调节饮品甜度、温度，满足消费者的个性化需求。在设计上，一些茶企（如茶颜悦色）巧妙地将传统中式元素与现代简约风格相结合，打造出既保留中式韵味又体现现代美感的产品。在场景营造上，一些茶馆（如上海的隐溪）打破了传统茶馆陈旧、老派、慢节奏的刻板印象，将场馆打造成兼具情调和腔调的时尚社交场所。

In terms of flavor, new-style tea drinks push the boundaries by incorporating a variety of fruits and ingredients, achieving innovative "crossovers" with unique formulations. They also allow customization of sweetness and temperature to cater to individual tastes. In design, brands like Chayan Yuese skillfully merge traditional Chinese aesthetics with modern minimalist styles, creating products that retain the essence of Chinese tea culture while embracing contemporary elegance. In ambiance, tea houses such as Yinxi in Shanghai have redefined the traditional tea-drinking experience. Moving away from the outdated, slow-paced image, they have transformed their spaces into stylish social hubs that blend sophistication with a trendy, inviting atmosphere.

一杯中国茶，一段中国缘
Sharing the Story of Chinese Tea with the World

在茶产业的未来发展中，茶将在健康、社交、文化、经济、时尚等领域展现更广阔的前景，并迎来更加多元的发展机遇。

As the tea industry continues to evolve, tea is set to unlock even broader potential across health, social, cultural, economic, and fashion sectors, embracing more diverse opportunities for growth.

中国是茶的故乡，作为爱茶人的我们，理应肩负起讲好中国茶故事的使命。通过展现中国茶的多重属性，推动茶文化的创新与全球传播，让世界更深入地了解中国茶的独特魅力，让更多人因茶而了解中国，因茶而爱上中国，因茶而走进中国。

China is the birthplace of tea. As Chinese tea enthusiasts, we carry the responsibility of telling the story of Chinese tea to the world. We must showcase the various dimensions of the tea, drive tea innovation, and promote global cultural exchange. By doing so, we can deepen the world's appreciation for the unique charm of Chinese tea, inspiring more people to explore China through tea, embrace China because of tea, and come to China because of tea.

第一部分
茶出中国
Tea's Chinese Origins

01 茶之源
The Origin of Tea and Its Development

1.1 茶的起源 / The Origin of Tea

"开门七件事，柴米油盐酱醋茶"，茶是寻常百姓居家过日子的必备之物；"琴棋书画诗酒茶"，茶亦是文人雅士闲情逸致的载体。在历史长河中，大俗大雅的茶虽然从来都不是主角，但让中华文化散发出沁人心脾的清香。

As a popular Chinese saying goes, a day starts with seven daily necessities: firewood, rice, oil, salt, sauce, vinegar and tea. From this saying, we can see the importance of tea in Chinese daily life. While for intellectuals, tea is part of graceful life, in the same league as music, chess, calligraphy, painting, poetry and wine. Tea suits both refined and popular tastes. Throughout its long history, tea has been an essential part of Chinese culture.

茶

（1）茶树原产地 / The Origin of the Tea Plant

中国是茶树原产地，是茶的故乡；中国是最早利用和栽培茶的国家，是茶文化的发源地。早期人们认为茶树原产于中国，但自 1824 年英国人罗伯特·布鲁斯在印度阿萨姆发现野生茶树后，这一观点便产生了分歧。今天，茶树原产地仍是学术界有争议的问题之一。

The tea plant, a native of China, was known to Chinese from very early times. As the first country which discovered, consumed and cultivated tea thousands of years ago, China is the birthplace of tea culture. However, the origin of the tea plant has remained a disputed issue. For a long time, people believed tea was unique to China. In 1824, wild tea trees were discovered in Assam area of India by a man named Robert Bruce from the UK, which led to the dispute of the origin of the tea plant.

野生大茶树是确定茶树原产地的重要依据之一，但仅凭这一点就认定印度是茶树原产地是远远不够的。在中国和印度均发现野生大茶树，但中国的野生大茶树数量之多、树龄之长、分布之广，非印度可比；当印度人还不知茶为何物时，中国人发现并利用茶就已有数千年历史；此外，印度为茶树原产地的说法没有史实支撑[①]，而中国为茶树原产地的观点不仅有文献记载，更有实物支撑[②]。由此，在这一问题上，

---

① 有人认为茶是印度史诗《罗摩衍那》（*Ramayana*）中提到的"神秘"药材，但此说法仅限于猜想。此外，印度茶树种植的起始是外源性的，该国最早也是最主要的产茶区阿萨姆的土著居民没有植茶、制茶和饮茶的历史。

② 出土于浙江余姚河姆渡文化田螺山遗址的古茶树根遗存证实中国人工栽培茶树的时间在距今 6000 年前后。出土于山东邹城邾国故城战国墓的茶碗与茶叶遗存是考古发现年代最早的饮茶实物证据，距今约 2400 年。此外，出自 2000 多年前汉景帝陵墓的茶叶实物也具备重要的考古价值。我国乃至全世界最早关于饮茶、买茶的记载是西汉文学家王褒所写的《僮约》。

虽然学术界存在不同看法，但人们普遍认为中国是茶树原产地。

The existence of wild tea trees is the most important criterion for the origin of the tea plant. However, it is nowhere near enough. So far, wild tea trees found in China heavily outnumber those in India and they are much older and taller. And the geographical distribution of wild tea trees in China is also wider than that in India. While the claim of India-tea-origin cannot be supported by any historical facts—no record of any kind, China-tea-origin is corroborated by ancient legends, books and physical evidence. Therefore, China is widely believed to be the origin of the tea plant.

人工种植茶树根

（2）茶之初现 / The Discovery of Tea

关于茶的起源，主要有两种说法，一是"药物说"[①]，二是"食物说"。很多人认为，茶最初被当作药物，主要源于"神农尝百草，日遇七十二毒，得荼而解之"[②]的记载。这个广为流传的故事是我国最早发现和利用茶叶的记载。另一种说法或许更为合理，即原始人在采集和尝试植物作为食物的过程中，偶然发现了茶，在食用茶的过程中，

---

① 日本学者冈仓天心用英文撰写的 *The Book of Tea*（《茶之书》）于1906年在纽约出版，该书堪称茶文化"东学西渐"第一书。该书的开篇之语即为 "Tea began as a medicine and grew into a beverage"（茶初为药品，后为饮品）。

② 这是世界范围内最广为人知的茶的传说。国外诸多知名茶叶品牌如立顿、川宁等均以此传说为茶的起源。

神农尝百草

逐渐认识了茶的药用功能。

Opinions vary regarding the discovery of tea, with two prevailing theories: "the food theory" and "the medicine theory". The most popular legend about the discovery of tea tells us that "an ancient tribal ruler named Shennong tasted lots of herbs in his search for ones with medicinal properties. He once tested 72 herbs in a single day and got poisoned. He was cured by tea which was discovered by chance."

Perhaps a more reasonable explanation of the discovery of tea is that primitive men found the therapeutic properties of tea leaves by chance while they were collecting and testing herbs for food.

（3）"茶"字的变迁与意义 / The Life Story of the Word "Tea" and Its Connotation

在古代史书中，茶的名称很多，如荼、荈、槚、茗等[1]。中唐时的陆羽在《茶经》中规范了茶的读音及书写——茶（Cha），从而结束了对茶称呼混乱不清的历史。"茶"字的构词很有意思，拆开来看就是"人"在"草（艹）""木"间，这样的妙解，展示出了中华文化中人与自然和谐共处的理念。

---

[1] 荼是最早的茶字，被认为是茶的古体字，最早出现于《诗经》，如"谁谓荼苦，其甘如荠"。在唐玄宗撰写的《开元文字音义》中，将含义为茶的"荼"改为"茶"，后经陆羽的《茶经》得以"推广"，沿用至今。茶还有一些有趣的别称，如嘉草、甘露、不夜侯、涤烦子、余甘氏、苦口师、忘忧草等。

Tea is alluded to in ancient classics under various names such as Tu, Chuan, Jia, and Ming, etc. Lu Yu in the middle Tang Dynasty (in the 8th century) standardized the pronunciation and writing of the Chinese character tea in *The Classics on Tea*[①], since then the ideograph 茶 (Cha) has been widely accepted. The connotation of the Chinese character of tea (茶) is an interesting reflection of Chinese philosophy of "the Harmony of Nature and Humanity", which is formed with Chinese character people (人) between grass (草 / 艹) and trees (木).

人在草木间（静观　供图）

---

① *The Book of Tea* 中将《茶经》译为 *The Holy Scripture of Tea*。

1.2 茶文化的形成与发展 / The Formation and Development of Chinese Tea Culture

（1）魏晋南北朝之前（220 年前）/ Before the Wei, Jin, Southern and Northern Dynasties (Before 220AD)

中国人对茶的利用历史悠久，巴蜀地区（中国西南部）是茶的摇篮。据古书记载，巴蜀人早在 3000 多年前就种植茶叶，并将茶叶作为贡品献给王室[①]。

Chinese people have been using tea for thousands of years, with Bashu (southwest China, area between Sichuan and Yunnan provinces) being the cradle of tea. According to ancient books, more than 3000 years ago, people in Bashu planted tea and offered it as a tribute to the royal family.

在 2000 年多前的西汉时期，茶已成为商品[②]，饮茶的习俗主要局限于巴蜀，而彼时的成都是茶叶消费的中心，也可能是最早的茶叶集散中心。

Tea had been a commodity since the Western Han Dynasty (202BC – 8AD) about 2000 years ago when the custom of drinking tea was mainly limited to Bashu area. At that time, Chengdu was the center of tea consumption and it might be the earliest tea distribution center as well.

---

① 据《华阳国志》记载，公元前 1046 年，周武王伐纣时，巴蜀以茶"纳贡"，这是茶作为贡品的最早记述。

② 王褒的《僮约》是一篇买卖奴婢的契约，成书于公元前 59 年。其中有"烹茶尽具""武阳买茶"的描述。前一句反映西汉成都一带，富人家中已出现饮茶用具；后一句反映茶叶已经商品化，还出现如"武阳"一类的茶叶市场。这是茶作为商品交易的最早记录。

巴茶贡周

王褒《僮约》

（2）魏晋南北朝（220—589）/ The Wei, Jin, Southern and Northern Dynasties (220–589AD)

魏晋南北朝时期，饮茶文化进一步普及。随着文人饮茶风气兴起，茶的食用方式开始从茶粥、茶羹向饮品转变，逐渐规范化。这一时期，奢靡之风盛行，有识之士以茶倡廉示俭，赋予茶伦理美德；随着道教、佛教的兴起和发展，茶进入宗教领域，成为精神饮品[①]；文人开启以诗赋形式赞颂茶的先河[②]，这些不仅赋予茶更多精神上的意义，也成为茶文化萌芽的标志。

茶禅一味（高锋 供图）

---

① 魏晋时期，受道教影响，人们认为茶使人轻身换骨，有利于养生，有助于修仙；南北朝时，佛教盛行，饮茶有助于坐禅时保持神思清明，"茶禅一味"思想产生。

② 西晋（265—317）杜育的《荈赋》描绘了茶叶从种植到品饮的全过程。《荈赋》是目前所知最早的专门歌吟茶事的诗赋类作品。

The Wei, Jin, Southern and Northern Dynasties saw increasing popularity of tea. The way of processing tea changed and apart from making tea into tea porridge or soup, people began to drink tea as a beverage. During that period, crazy aspiration of competing wealth and luxurious lifestyle was prevailing, and in order to discourage such a social trend, some enlightened people used tea as a substitute for liquor to promote integrity and incorruptibility, thus bestowing tea a more spiritual significance. With the development of Taoism and Buddhism, tea was consumed and cultivated by more religious people; meanwhile scholars began to praise tea in literary works. As a result, tea was attached with spiritual value, which marked the dawn of tea culture.

（3）隋唐时期（581—907）/ The Sui Dynasty (581–618AD) and the Tang Dynasty (618–907AD)

在隋代不足 40 年的历史中，有关茶的记载不多，但隋朝统一全国并修筑了贯通南北的大运河，这对于促进唐代的经济文化发展起到铺垫作用。

In the Sui Dynasty, which lasted for less than 40 years, there were few records about tea. But a unified country was established and the Grand Canal connecting the north and the south was constructed, which paved the way for the economic and cultural prosperity of the Tang Dynasty.

唐代是中国茶文化正式形成时期。盛世大唐为茶叶的生产与流通以及茶文化的发展创造了有利条件。茶事活动不仅受到皇家和寺院的推动，也受到文人士大夫的影响[①]。这一时期，茶叶的生产、消费与贸

[①] 皇家给茶业发展提供了制度保障；佛教禅宗对茶文化进行宣传推广；文人咏茶进一步丰富了茶文化的内涵。

大唐贡茶院（龚记永元　供图）

易相互促进，茶的饮用得到进一步推广，成为举国之饮；完备的贡茶制度①、茶税制度②、茶马交易制度逐渐形成③；茶文学繁盛④，世界上最早的茶学专著《茶经》问世。该书系统阐述了茶及其相关知识与实践，标志着中国茶文化的正式形成。

Chinese tea culture was formally formed in the Tang Dynasty. The prosperous Tang Dynasty provided favorable atmosphere for the production and circulation of tea and the development of tea culture. The royal family, temples of Buddhism and Taoism and literati all contributed to the popularity of tea. With the stimulation of tea production, consumption and trade, tea

---

①　贡茶在唐代进入制度化发展阶段，贡茶地与贡茶数量倍增。建于唐代宗大历五年（770）的顾渚贡茶院开古代官办茶焙之先河。唐代以官营贡焙为主的贡茶制度被宋代继承，极大地促进了茶文化的发展。

②　唐建中元年（780），诏征天下茶税，十取其一，是为茶税之始。

③　如《新唐书·陆羽传》记载"时回纥入朝，始驱马市茶"。唐代封演所著《封氏闻见记》卷六《饮茶》也有记录说"往年回鹘入朝，大驱名马，市茶而归，亦足怪焉"。

④　茶文学早在唐代之前就已出现，中晚唐以后发展繁荣，形式包括茶书、茶诗和茶文。据统计，《全唐诗》中与茶相关的诗有 600 多首。

spread across the country and became the national drink. Tribute-tea system, tea-tax system and tea-horse-trade system were established. The world's earliest treatise on tea *The Classics on Tea*, in which knowledge of tea and related practices were elaborated systematically, marking the formation of Chinese tea culture.

唐代煮茶（静观 供图）

（4）宋元时期（960—1368）/ The Song Dynasty (960–1279AD) and the Yuan Dynasty (1271–1368AD)

茶兴于唐，而盛于宋。宋代是茶文化发展的黄金时期。宋代皇室多醉心于茶，形成豪华极致的宫廷茶文化；文人雅士相邀饮茶，各类茶会、茶宴精彩纷呈，促进了茶与其他艺术形式的融合。同时，茶进一步普及成为日常生活必需品，催生了趣味横生的民间茶文化。宋改煎煮茶为点茶①。分茶是宋代流行的一种茶艺，简而言之，在点茶的基础上于茶面作画。宋代最有趣的茶俗当属斗茶，斗的就是点茶技艺的高低、茶叶品质的高下，有时也比拼分茶技艺。斗茶茶汤以白取胜，

---

① 点茶是宋代生活美学的体现。完整的点茶有大约15个步骤，常用工具多达12种。

茶盏尚黑，故黑釉盏 ① 尤其是建盏最受推崇。

The Song Dynasty was a golden age of tea. The imperial family of the Song Dynasty were obsessed with tea, boosting the luxurious royal tea culture; refined

宋代点茶沫饽（静观 供图）

scholars promoted the integration of tea and other forms of art by holding tea parties or banquets; meanwhile, tea became a necessity of daily life, which promoted the interesting folk tea culture. In this period, the way of tea preparation changed from boiling to whipping, which led to the popularity of tea competition among all classes of society. Besides, there was a special tea art called "Fen Cha" or "Cha Danqing" , literarily meaning "tea painting"—painting on the foam of tea with water and the whisk. Since the key to win tea competition was the color of the foam—the whiter, the better, black glazed tea bowls especially those made in Jian Kiln (now at Jianyang district of Nanping city, Fujian province) were highly praised.

　　宋代茶类增多，名茶辈出；饼茶是主流茶类，但散茶（也称草茶）逐渐风行，花茶也开始出现。在制茶技艺上，团茶、饼茶制茶工艺愈加精细，达到历史高峰。这一时期，东南茶区（主要指今天的福建和浙江一带）已成为茶叶生产和技术中心。

With highly-developed processing technology of tea cake, there were a great number of famous teas. The tea cake was the major form of tea

---

　　① 黑釉盏能更好地衬托出白色茶汤和泡沫。黑釉盏中最著名的是产自建窑（位于今福建省南平市建阳区）的建盏。建盏的釉色以兔毫斑、鹧鸪斑、油滴、曜变等为代表，斑纹为烧窑过程中气温火候变化所致，每一件都独一无二。

宋代建窑兔毫盏（内蒙古博物院藏）

products but loose tea (also called Cao Cha tea, or roasted tea leaves) began to gain increasing popularity. In the Song Dynasty, scented tea started to appear. The center of tea production had moved to the southeast region (mainly now Fujian and Zhejiang provinces).

在饮茶风气的普及①、城市商业经济高度繁荣的大背景下，宋代茶馆兴盛②，茶馆文化发展。以茶为题材的艺术作品不胜枚举，尤以散文、诗词为佳。

In the Song Dynasty, with the popularization of tea drinking, teahouses could be found everywhere. There were many literary works on tea, especially in the form of prose, poetry and lyrics.

元代是中国历史上首次由少数民族建立的大一统王朝，作为元代统治民族的蒙古族，以奶制品、肉类为主要食物，需要饮茶助消化，故对茶叶的生产比较重视；同时他们将游牧地区的饮茶习俗，即茶中加盐加奶的方式，带入中原地区。

The Yuan Dynasty was the first unified dynasty established by the ethnic minority in Chinese history. As the ruling ethnic group of the Yuan Dynasty, the Mongolian people mainly consumed milk products and meat, so they needed to drink tea to promote digestion. Therefore, they attached much importance to the production of tea; meanwhile, they also brought the

---

① 王安石在《议茶法》中说："夫茶之为民用，等于米盐，不可一日以无。"《梦粱录》卷十六《鲞铺》载："盖人家每日不可阙者，柴米油盐酱醋茶。"

② 我国茶馆成形于唐代。唐宋时称茶肆、茶坊、茶楼、茶邸，明以后始称茶馆。

tea drinking customs of nomadic areas, by adding salt and milk into tea, to the Central Plains region.

茶的生产和饮用方式虽然基本沿袭宋制，但茶的制作、品饮和价值取向均处于转变期。除了供皇室贵族享用的团饼茶外，散茶成为主要生产茶类，在民间普及。

The processing technology, drinking methods and spiritual value of tea in this period underwent gradual transformation. In addition to the tea cake used by the royal family and nobles, loose tea was consumed by more and more ordinary people.

从某种程度上说，元代统治者对南宋精细奢靡的品茶之风并不认同，对茶的态度主要是饮食习惯的接受。因此元代既没有产生有影响力的茶学专著，也很少出现涉及茶的诗词、文章，从这个意义上说，元代的茶业、茶事有所消淡[①]。

To some extent, tea fell out of favor as a drink during the years of the Mongol Yuan Dynasty when the Mongolian rulers considered the deliberate drinking of tea of Song Dynasty a symbol of decadence. Therefore, although tea had been an indispensable part for the people in the Yuan Dynasty, there were very few poems or articles involving tea in that period, let alone tea monograph. In a sense, there was retrogression in tea industry and its culture in the Yuan Dynasty.

---

① 元代，茶事虽消淡，但因在武夷山设置御茶园而开启武夷茶的荣光。1302年，元在武夷山九曲溪设御茶园，武夷茶兴起，此后数百年，"茶必武夷"。

（5）明代（1368—1644）/ The Ming Dynasty (1368–1644AD)

紫砂壶鼻祖——供春壶

明代是中国制茶工艺与饮茶方式发生重要变革的阶段，也是我国茶文化发展的另一个高峰期。1391 年，明代开国皇帝下诏"废团改散"，即用散茶代替团饼茶进贡。"废团改散"顺应了简化制茶工序、减少烘焙流程的社会需求。摆脱团饼茶的传统束缚，加上炒青工艺日渐成熟①，散茶制作工艺的创新达到历史巅峰，各地名茶辈出。随着人们对散茶加工技艺的不断追求与探索，其他茶类加工工艺逐渐形成并发展。制茶方式的变化必然改变饮茶方式。明人改用茶壶泡散茶，用沸水冲泡，简化了饮茶流程。饮茶方式的变化也促使以紫砂壶为代表的茶器具的变革。对饮茶环境和冲泡用水的讲究也是明代茶饮的特点和贡献②。在茶业蓬勃发展的大背景下，明代文人纷纷以书、画等形式表达他们对茶的理解与喜爱。明代茶书众多，现存 50 余种，约占现存中国古代茶书总数的一半。

Tea experienced a renaissance in the Ming Dynasty. In 1391, the founding emperor of the Ming Dynasty ordered loose leaf tea to be offered as tribute instead of traditional tea cake. Thanks to this imperial decision and the popularity of the frying pan, technology in Chaoqing tea, or pan-fried green tea, gradually matured and loose leaf tea was booming. Meanwhile, more varieties of tea were invented or created due to the innovation in tea

---

① 唐宋以蒸青为主，炒青技术开始萌发。唐代刘禹锡《西山兰若试茶歌》中有"斯须炒成满室香"的诗句，又有"自摘至煎俄顷余"之句，说明嫩叶经炒制后生香，且炒制花费的时间不长。这是目前发现的关于炒青绿茶最早的文字记载。

② 有学者认为明中期后的茶事是中国茶事的"文艺复兴"。

processing technology. It was in this period that tea began to be brewed by steeping loose leaf tea in boiling water. As a result, there were great changes in tea utensils, and consequently purple clay teapots got popular. Besides, people in the Ming Dynasty attached great importance to the surrounding or atmosphere when drinking tea and they were very fastidious about brewing water. In addition, there are over 50 tea books written in the Ming Dynasty, accounting for half of the extant ancient Chinese tea books.

（6）清代（1616—1911）/ The Qing Dynasty (1636–1911AD)

清代，茶文化在局部地区继续发展，但总体上呈现由盛转衰之势。茶与人们日常生活的结合更加紧密，六大茶类均已产生。清代与明代饮茶方式大致相同。宫廷饮茶带有北方游牧民族的影子[①]，且在器具使用上更为讲究[②]。

In the Qing Dynasty, tea industry experienced the cycle from prosperity to decay. It kept developing in some regions, with all six basic teas being created. It played a more comprehensive role in people's daily life. Generally, there was little change in the way of tea drinking. While in the imperial palace, tea drinking was influenced by the tea custom of the nomads from the north, since the Qing Dynasty was founded by Manchu Nationality. Besides, tea ware used by the royal family was more exquisite.

[①]　饮用奶茶成为宫廷茶事的重要组成部分，清宫中的奶茶也叫"奶子茶"，是满族的传统饮品。其制作时在取奶、选茶、配器、用水方面有诸多讲究。除日常饮用外，饮用奶茶也成为国家礼仪制度的一部分。

[②]　如名贵的宫廷御用珐琅彩茶具。珐琅彩瓷器初创于康熙晚期，盛烧于雍正、乾隆年间。

清代名茶

19世纪80年代末，中国茶业在印度、斯里兰卡、日本等洋茶围剿下，已呈江河日下之势，茶叶出口量开始连年下降。

Since the late 1880s, due to the strong competition of teas from India, Sri Lanka and Japan, exports of Chinese tea kept declining.

清代末期至民国初期是旧中国茶馆业的巅峰时期。茶馆的经营方式和功能更加多元化，它将茶、曲艺、戏剧、地方美食等多种元素融为一体，成为具备休闲娱乐、止渴充饥、信息传播、社交会友、行业聚会、市场交换、仲裁审判等功能的公共空间。

The period from the late Qing Dynasty to the early Republican Period was the heyday of Chinese teahouse industry. At that time, the teahouse integrated tea with many other elements such as folk art, Chinese operas and local cuisine. It was virtually a multifunctional public place for thirst quenching, leisure and entertainment seeking, information exchanging, social and industrial gathering, even arbitration and trial.

清代茶叶外销

（7）民国时期（1912—1949）/ The Republican Period (1912–1949)

民国时期，中国社会动荡、战争频仍。1921 年，中国茶产业发展达到历史谷底[①]。曾经一枝独秀、经历了几个世纪全球市场垄断的中国茶在凄风苦雨中蹒跚前行。20 世纪初，民国政府派出最早一批留学生赴日本、印度、斯里兰卡等国，考察学习当地茶叶情况。这批学生学成归国后，开始在国内各大茶区兴办茶叶实验场及研究所。他们的付出和努力影响了百年后中国茶产业的布局。

Undergoing much suffering during the Republican Period, China had a hard time and so did Chinese tea industry. In 1921, Chinese tea industry hit rock bottom in history, with both the international and domestic market

---

[①] 1921年，中国茶叶出口量跌至2.6025万吨。中国茶叶从昔日全球超90%的出口占有率，萎缩至仅 8.79%。不仅外销市场衰落，内销市场也被国外茶叶侵占，中国茶遭遇前所未有的冰点。

shares being totally lost. For the first time, Chinese tea was forced to reassess its survival. At the beginning of the 20th century, the government sent a batch of students to Japan, India and Sri Lanka to learn new technology of tea. After returning to China, they devoted themselves to the tea industry by setting up tea experimental plantations and research institutes. Their earnest efforts paved the way for the development of Chinese tea industry in the future.

清末茶馆

（8）新中国成立之后（1949 年至今）/ After the Founding of the PRC (From 1949 to the Present)

　　新中国成立后不久，中国茶产业开始恢复，特别是改革开放之后，国家采取了一系列措施助力茶业复兴[①]。到 21 世纪初，我国茶叶生产进入全面发展时期，无论在茶叶种植面积、茶叶产量还是茶消费总量、

　　[①]　20世纪60年代，南茶北移进山东，东茶西扩至甘肃、西藏等措施使中国茶区进一步扩大。

出口金额等方面都位于世界前列[①]。今天，茶的各方面研究和开发已见成效。茶叶科研体系基本建成，茶学教育不断完善，茶产业链不断延伸[②]，茶文化事业欣欣向荣，"茶为国饮"理念深入人心。

After the founding of the PRC in 1949, especially after the reform and opening up, a series of measures have been taken to revive the tea industry. After decades of efforts, the new century saw encouraging signs of the revival, with total planting area, tea production and consumption, and export volume among the highest in the world. Today, Tea science and researches have made tremendous gains. As the most popular beverage in China, tea has been the national drink of China.

传统制茶（静观　供图）

2019 年 11 月，在中国的倡议和推动下，联合国宣布将每年 5 月 21 日确定为"国际茶日"[③]。2022 年 11 月，"中国传统制茶技艺及其相

---

① 2023 年，中国茶叶种植面积、产量、消费总量均居世界第一。

② 目前，茶的有效成分被广泛应用于药物、保健品、功能食品及化妆品中，在延伸产业链的同时，提升了茶的附加值。

③ "国际茶日"是中国首次成功推动设立的农业领域国际性节日。2019 年 6 月，中国茶叶学会提出设立"国际茶日"提案，在联合国粮食及农业组织大会第 41 届会议上审议通过，并提交给联合国大会。

关习俗"被列入联合国教科文组织人类非物质文化遗产代表作名录。这是关于茶园管理、茶叶采摘、茶的手工制作，以及茶的饮用和分享的知识、技艺和实践，对于弘扬中国茶文化、深化文明交流互鉴具有重要意义。

Chinese tea is regaining its former glory in the world. In November 2019, by China's suggestion and promotion, the United Nations declared May 21st as International Tea Day. In November, 2022, "traditional tea processing techniques and associated social practices in China" was inscribed on the UNESCO's Representative List of the Intangible Cultural Heritage of Humanity. It consists of knowledge, skills and practices concerning management of tea plantations, the picking of tea leaves, and the processing, drinking and sharing of tea. It is of great significance for promoting Chinese tea culture and deepening cultural exchange and mutual understanding.

今天，中国茶正以积极的姿态展现其年轻化、国际化、科技化的一面。茶业成为联动第一产业、第二产业、第三产业的朝阳产业。它与乡村振兴、精准扶贫、"一带一路"、健康中国、讲好中国故事等国家政策或倡议高度相关。

As tea industry has realized the importance of young consumers and international market, it is trying to create an energetic and modern image worldwide; meanwhile it is committed to the science and innovation of tea. Today, as a promising industry, tea industry links the primary, secondary, and tertiary industries. It is highly related to national policies or strategies such as rural revitalization, targeted poverty alleviation, the Belt and Road, healthy China, and telling China's stories well.

机器制茶（静观　供图）

　　回望茶数千年的历史，种茶技术不断提升，备茶方法几度变化，茶从最初的药用、食用物品，发展到今天的大众饮品。茶始终在国家政治、经济和文化生活中扮演重要的角色。可以说，茶凝聚了中华民族的智慧、传播了中华文化、影响了历史的进程。这片树叶是我们宝贵的遗产，是我们灿烂文化的名片。

　　Over the millennia, both cultivation techniques and methods of tea production have seen numerous advances. Tea has evolved from a medicinal and edible plant to a favorite beverage up to this day, playing a significant role in China's politics, economy and cultural exchanges. It is the epitome of Chinese culture and wisdom, and it has touched and changed our lives as no other beverage has. As precious part of our national heritage, it is one of the best spokesmen of China.

小贴士：

　　1. 在介绍宋代"茶百戏"时，可将"茶百戏"类比为"咖啡拉花技艺"（Latte Art）。但需说明咖啡拉花是将打成奶泡的牛奶加入咖啡形成图案；而茶百戏则是用清水在茶汤表面变换图案，且因为茶汤的流动性，画面维持时间较短。2017年，"茶百戏"被列入福建省非物质文化遗产代表性项目名录。

2. 关于茶的发展历程，冈仓天心在 *The Book of Tea* 一书中的划分和描述非常有意思，原文和译文如下："Like art, tea has its periods and its schools. Its evolution may be roughly divided into three main stages: the Boiled Tea, the Whipped Tea, and the Steeped Tea… The Cake-tea which was boiled, the Powdered-tea which was whipped, the Leaf-tea which was steeped, mark the distinct emotional impulses of the Tang, the Song, and the Ming dynasties of China. If we were inclined to borrow the much-abused terminology of art-classification, we might designate them respectively, the Classic, the Romantic, and the Naturalistic schools of Tea."（"用来煎煮的茶饼，用来拂击的茶末，和用来淹泡的茶叶，分别鲜明地代表中国唐代、宋代，以及明代的感情悸动。再次让我们借用已经相当浮滥的美学术语，将它们挂上古典主义、浪漫主义与自然主义的流派之名。"）此段译文选自谷意的《茶之书》译本，冈仓天心以古典主义、浪漫主义和自然主义对应煎茶、点茶和泡茶，用不足百词的篇幅，勾画了中国茶及茶文化发展的主要阶段，且语言简练优美，便于西方读者理解，颇有借鉴价值。

3. 根据教育部 2022 年底公布的数据，中国有 40 多所中职和 80 多所高等院校开设茶学和茶文化专业，每年培养制茶、茶艺等专门人才 3000 多人。（According to the data China's Ministry of Education released in 2022, there are over 40 vocational colleges and 80 universities in China that have set up majors in tea science or tea culture, resulting in over 3,000 graduates specializing in tea production and art every year, according to the ministry.）

1.3 中国茶文化发展进程中的重要人物 / Big Names in the History of Chinese Tea Culture

（1）传奇神农氏 / The Legendary Emperor—Shennong

说起茶的起源，中国人首先会想到神农氏，一位生活在公元前3000 年左右的神话人物。作为中华民族人文始祖之一，神农利用神力造福人民，在中国人眼中，他是农业之神、医药之神，同时也被奉为茶祖[①]。

When we talk about Chinese tea culture, we will immediately think of the legendary Emperor, Shennong, who lived around 2700 BC. As one of the legendary ancestors of Chinese people, Shennong used his divine influence to benefit the ancient people. Regarded as God of agriculture and medicine, Shennong has been acclaimed as the ancestor of tea as well.

在 5000 多年前的新石器时代，随着人口数量增长，生产方式从狩猎采集向农耕转变，先民必须大量采集、辨别食物，而辨别的方式只能是口尝。据说，神农在煮水时，头顶的树叶飘落于锅中，机缘巧合下，神农发现了茶。虽然史书对"神农尝百草"亦有记载，但在 5000多年的历史长河中，神农的故事常被视为传说。更为合理的解释是"尝百草"不是个人行为，而是中国一代又一代先民的群体行为；相应地，茶的发现亦是如此。

About 5000 years ago, in the Neolithic age, with the growth of

---

① 陆羽《茶经》的"六之饮"篇载："茶之为饮，发乎神农氏，闻于鲁周公。"这是目前有据可考的最早确定神农氏茶祖地位的文献。

population and change of lifestyles from hunting and gathering to farming, primitive people had to gather and test various plants or herbs. They had no better ways of testing but to eat them by themselves. The ancient legend told us that Shennong found this new drink when one day a few leaves from an overhanging tree fell into his pot of boiling water. With a history dating back about 5000 years, it was inevitable that the legend steeped into storytelling. A more reasonable explanation is that the discovery of tea was not an individual behavior, but a contribution made by generations of primitive people.

（2）茶圣——陆羽 / The Tea Sage—Lu Yu

8 世纪中叶，第一本茶学专著问世，这就是唐代陆羽（733—804）的《茶经》。陆羽生活在儒、释、道三教共存共荣的时代，他在茶事中发现了存于世间万物的和谐之道。《茶经》构建了茶的基本准则，全书共三卷十章，对种茶、制茶、品茶等进行了系统而全面的总结。

The world's earliest treatise on tea *The Classics on Tea* was written by Lu Yu (733–804) in the middle of the 8th century. Lu Yu was born in an age when Buddhism, Taoism, and Confucianism were seeking mutual synthesis. He saw in the tea the same harmony and order which reigned through all things. In the book he formulated the code of tea. Consisting of three volumes and ten chapters, the book discussed all known information about tea, such as the nature of tea-plant, tea origin, variety, distribution, cooking, utensils and tea spirit.

尤其值得一提的是，在"四之器"中，陆羽列出了唐时茶事使用的二十几种器具，其完备程度足以让今天最讲究的茶事相形见绌；在

"五之煮"中，陆羽提出烹茶要舍弃除了盐之外的所有调料，同时也论述了水的选择和水温的把握。在随后的章节中，陆羽讨论了彼时饮茶的粗鄙之处、历史上的著名茶人、知名茶区、饮茶的不同方法以及图示等。至此，饮茶从解渴式的粗放型饮法向细煎慢啜的品饮型饮法过渡，成为一种高雅的艺术活动。

In Chapter Four, Lu Yu described the implements needed for the preparation of tea—more than 20 in all, a number which would put even the most elaborate modern tea service to shame! In the fifth chapter, he described the method of cooking tea. He eliminated all ingredients except salt and he also dwelled on the much-discussed question of the choice of water and the degree of boiling it. In the following chapters, he talked about the popular but vulgar ways of tea drinking and listed a historical summary of illustrious tea-drinkers, famous tea plantations, the possible variations of the tea-service and illustrations of the tea-utensils. For the first time, tea drinking was endowed with spiritual and humanistic connotation.

此外，陆羽还首次赋予饮茶精神境界和人文内涵①。茶文化作为独立的文化分支建立起来。《茶经》将饮茶从物质生活需求上升为精神生活需求，被视为茶文化发展历程中里程碑式的作品。因此，陆羽被

陆羽像

---

① 《茶经·一之源》载："茶之为用，味至寒，为饮，最宜精行俭德之人。"此后，"精行俭德"四字成为历代茶人追求的目标。

《茶经》（苏泓铭　供图）

称为"茶圣"[①]，并被尊为茶业之鼻祖。

    Since then, tea drinking has been promoted as a graceful and artistic practice. So influential was *The Classic on Tea* in spreading tea drinking throughout the country that Lu Yu was worshipped as "Tea Sage" (or "Tea Saint") and "Father of Tea Industry".

（3）茶僧——皎然 / The Tea Monk—Jiao Ran

    历史上第一个将"茶"与"道"联系在一起的是唐代茶僧皎然（约760—840）。在《饮茶歌诮崔石使君》一诗中，皎然提出了"孰

---

    ① 后世茶馆常供奉陆羽瓷偶，以求生意兴隆。如欧阳修曾在《集古录跋卷》中写道："茶载前史，自魏晋以来有之。而后世言茶者，必本鸿渐，盖为茶著书，自羽始也。至今俚俗卖茶，肆中多置一瓷偶人，云是陆鸿渐。至饮茶客稀，则以茶沃此偶人，祝其利市，其以茶自名久矣。"

知茶道全尔真，唯有丹丘得如此"的道理，言下之意，茶道是诗者个人的追求，而只有少数人才能明了个中真义。除了提出"茶道"这个概念，皎然在诗中还描述了"一饮涤昏寐，再饮清我神，三饮便得道"①的品饮意境，即一碗茗茶需三饮，一饮神清气爽、二饮除却杂念、三饮参悟得道。因为对茶道的参悟，以及其在茶学、佛学和文学领域的造诣，皎然被视为中国禅宗茶道②的开创者。

Nowadays, Teaism (chá dào in Chinese) is familiar to many people, and the name itself reflects the close relationship between tea and "Dao" (frequently pronounced as Tao, an important word in Taoism). It is Jiao Ran (760–840), a famous tea monk of the Tang Dynasty, who first associated tea with "Dao", hence the term Teaism. In one of his poems, Jiao Ran proposed the term Teaism by saying "Dao" was the ultimate pursuit of tea drinking, while only a few people could understand the true meaning of Teaism then. In addition to putting forward the concept of Teaism, Jiao Ran also described the artistic mood of tea drinking. He said "the first bowl of tea gets one refreshed; the second bowl makes one carefree and the third cultivates wisdom of life". Because of his contribution in Teaism, Buddhism and literature, Jiao Ran is regarded as the founder of Chinese Zen Tea School.

此外，同为爱茶人，皎然与陆羽互为茶道师友，他是唐代诗人中咏及陆羽且现今存诗最多的一位，为后代研究陆羽提供了重要的资料。

In addition, as a fellow tea master and close friend of Lu Yu, Jiaoran,

---

① 原文为"一饮涤昏寐，情来朗爽满天地。再饮清我神，忽如飞雨洒轻尘。三饮便得道，何须苦心破烦恼。"

② 有文章提出，中国四大茶道分别为：贵族茶道，生发于茶之品，旨在夸示富贵；雅士茶道，生发于茶之韵，旨在艺术欣赏；禅宗茶道，生发于茶之德，旨在参禅悟道；世俗茶道，生发于茶之味，旨在享乐人生。

holds distinction as the Tang poet with the most surviving verses dedicated to the legendary Tea Sage. His works serve as invaluable primary sources for scholarly research on Lu Yu's life and legacy.

（4）亚圣卢仝 / Lu Tong, the Tea Sage Only Second to Lu Yu

中国不仅是茶之国，也是诗之国。茶带着风情雅致，步入诗歌的殿堂。唐代爱茶的诗人颇多，如白居易、元稹、皮日休、卢仝等。他们的诗作记述茶事之盛，诗的广泛流传促进了茶的传播与普及。

China is not only the country of tea, but also a country of poetry. In the Tang Dynasty, tea entered the realm of poetry as one of the polite amusements. Many poets such as Bai Juyi, Yuan Zhen, Pi Rixiu, and Lu Tong were famous tea lovers. They described the prosperity and elegancy of tea and its culture, which promoted the consumption of tea in a graceful way.

在古代数千首茶诗中，最广为人知的莫过于卢仝（约795—835）的《走笔谢孟谏议寄新茶》。该诗精华部分是诗人用排比句法，描述不同层次的饮茶境界，展现品茶得道的绝妙。"一碗喉吻润，两碗破孤闷。三碗搜枯肠，惟有文字五千卷。四碗发轻汗，平生不平事，尽向毛孔散。五碗肌骨清，六碗通仙灵。七碗吃不得也，唯觉两腋习习清风生。"这段也被称为《七碗茶歌》，广为流传，被历代文人雅士引用、化用，为中国茶诗之最。卢仝，自号玉川子，因精深的品茶造诣及绝妙的茶诗，被尊为"亚圣"和"茶仙"[①]。据说卢仝在日本被视为煎茶道

---

① "玉川煎茶"成为经典诗画题材。如苏轼曾写道："何须魏帝一丸药，且尽卢仝七碗茶。"宋代点茶全套器具称作"十二件大玉川先生"，也是以卢仝命名的。

的祖师爷，至今备受推崇^①。

Among thousands of ancient poems about tea, the most famous one is written by Lu Tong (about 795–835), a poet in the Tang Dynasty. When receiving freshly made tea from his friend Meng Jian, remonstrator of Emperor Xianzong of Tang, Lu Tong wrote a long poem. The essence of the poem reads "Drinking the first bowl of tea, I quench my thirst; with the second bowl, I no longer feel lonely; the third, I draw inspiration for writing; the fourth, I calmly face the reality; the fifth, I feel purified; the sixth bowl leads me to the realm of immortals; the seventh—oh, I could have no more, for I am flying to the fairyland." Lu Tong described seven levels of tea drinking so vividly that this part of the poem was also known as "*A poem on seven bowls of tea*". Mainly because of this poem, Lu Tong was considered one most important figure in Chinese tea culture only second to Lu Yu. In addition, held in high esteem in Japan, Lu Tong is honored the forefather of Japanese green tea ceremony.

卢仝像

---

① 据说在日本传统的煎茶道中，以喉吻润、破孤闷、搜枯肠、发轻汗、肌骨清、通仙灵、清风生的顺序，作为茶道的七道汤。

（5）"斗茶裁判长"——蔡襄 / The Referee on Tea Competition—Cai Xiang

生活在 11 世纪的北宋名臣、书法家蔡襄^①（1012—1067）创制贡茶极品"小龙团"，并撰写茶学专著《茶录》。《茶录》分上下两篇，细述彼时贡茶的重要产地建安北苑^②的茶品和茶器等，使建茶名垂天下，并为宋代艺术化的茶饮奠定了理论基础；书中关于茶的烹制的独特见解，如首次提出茶必须色、香、味俱全，描述斗茶的过程及明确判定胜负的标准等，使该书成为斗茶这一茶俗的指导性论著^③。

Cai Xiang (1012–1067), a famous official and calligrapher of the Northern Song Dynasty who lived in the 11th century, created the best tribute tea "Xiao long tuan " (a refined small tea cake with dragon and phoenix patterns on it) and wrote a book named *Record of Tea*. The book consisted of two chapters and detailed making and whipping of the tea cake produced by the imperial tea garden Beiyuan and tea bowls from Jian kiln. Thanks to his description and recommendation, tea produced in Jian' an of Beiyuan and tea bowls from Jian kiln were acknowledged as the best in the Song Dynasty. *Record of Tea* has long been regarded as the most important guiding book for tea competition.

---

① 蔡襄曾任福建路转运使，管理北苑贡茶生产，他与苏轼、黄庭坚、米芾并称"宋四家"。

② 北苑贡茶始于五代十国时期的闽国，兴于宋，又历经元、明二朝，历时458年，一直是贡茶中的上品。宋代把贡茶做到了极致，贡茶把建茶推向了辉煌。

③ 《茶录》中有"茶色贵白""茶味主于甘滑"等关于判定茶品质的描述；有"视其面色鲜白，著盏无水痕为绝佳"等关于斗茶判定标准的描述。

（6）误国遗恨、茶道精深——赵佶 / Reign's Ruin and Tea Mastery—Zhao Ji

宋徽宗赵佶（1082—1135）于12世纪最初的26年在位，这位"诸事皆能，独不能为君"的皇帝是一位茶学大师。其所著《茶论》[①]是唯一一本由皇帝撰写的茶叶著作。该书对北宋时期蒸青团茶的产地、采制、烹试、品质等均有详细记述，反映了北宋茶业的发达程度[②]。皇帝提倡、群臣趋奉，上行下效，因此宋代制茶之工艺益精，名茶辈出，斗茶之风日盛。此外，宋徽宗所绘《文会图》描绘了林间茶会的氛围和程式，是研究宋代茶文化的珍贵记录[③]。

Zhao Ji (1082–1135), Emperor Huizong of the Song Dynasty, reigning from 1100 to 1126, was too great an artist to be a well-behaved monarch. He wrote a book named *Treatise on Tea*, which is the only tea book written by an emperor. The book provided a panoramic view of the tea culture of the Northern Song Dynasty. The emperor's enthusiasm was flattered and followed by the whole society. As the result, techniques of processing tea were more sophisticated; tea competition got more popular; a great number of fine teas appeared. In addition, Emperor Huizong depicted a gathering of a group of scholars in the woods in his famous painting *Wen Hui Tu* (*Gathering of Scholars*), which offers precious evidence for the study of tea

---

① 因《茶论》成书于大观元年（1107），故后人称之为《大观茶论》。该书有地产、天时、采择、蒸压、制造、鉴辨、白茶、罗碾、盏、筅、瓶、杓、水、点、味、香、色、藏焙、品名、外焙等二十篇。该书和《茶录》等，共同构建了宋代饮茶的美学意境。

② 如宋徽宗在《茶论》中言及"采择之精，制作之工，品第之胜，烹点之妙，莫不咸造其极"。

③ 《文会图》绢本设色、立轴（Ink and colors on silk, Hanging scroll），纵184.4厘米，横123.9厘米，该图表现的是环桌而坐的文士正进行林间茶会。此图现藏于台北故宫博物院。

culture in the Song Dynasty.

蔡襄像

《文会图》（局部）

（7）慕茶诗客——苏轼等文坛大家 / Tea Loving Poets Represented by Su Shi

唐宋时期，不少文人墨客把饮茶视为一种净化心灵和寄托情思的活动，他们的诗词以文雅之句记述茶事之盛，诗词的广泛流传促进了茶的传播和普及。唐代著名诗人李白①、白居易②，宋代文豪欧阳修③、苏

---

① 李白以饮酒闻名，也曾写过茶文。如李白在获赠仙人掌茶后，作《答族侄僧中孚赠玉泉仙人掌茶并序》答谢。

② 白居易的诗句"坐酌泠泠水，看煎瑟瑟尘。无由持一碗，寄与爱茶人。"写出了烹茶品茶的意境。而《琵琶行》中"商人重利轻别离，前月浮梁买茶去"则为后人留下了重要的茶叶史料。

③ 欧阳修为蔡襄《茶录》作的后序中说："茶为物之至精，而小团又其精者，录序所谓上品龙茶是也。盖自君谟始造而岁供焉。仁宗尤所珍惜，虽辅相之臣，未尝辄赐。"这一段对团茶之贵重的描述极具史学价值。欧阳修还著有专论烹茶之水的《大明水记》。

轼、范仲淹^①、陆游^②等均是爱茶之人，也都留下了与茶相关的不朽的诗文。

During the Tang and Song Dynasties, many literati regarded tea drinking as an activity to purify the mind and relax the spirit. They drew their inspiration from tea drinking, and their beautiful works did much to popularize tea drinking. Famous literati in that period included Li Bai and Bai Juyi in the Tang Dynasty and Ouyang Xiu, Su Shi, Fan Zhongyan and Lu You in the Song Dynasty.

在众多大家之中，又以苏轼（1037—1101）为最。他熟知茶史，对品茶、烹茶都很擅长，不仅熟知茶的功效，还曾亲手种过茶，因而创作出不少既有文采又有见地的茶诗词。《叶嘉传》是苏轼以拟人化手法为茶叶所写的一篇传记，既是自喻，亦写出了茶人精神。而"从来佳茗似佳人""活水还须活火烹""何须魏帝一丸药，且尽卢仝七碗茶""酒困路长惟欲睡，日高人渴漫思茶"等诗句道尽了茶的魅力、烹茶用水之精髓及茶的日常功效等。

Among them, Su Shi (1037–1101) is the most famous one. Su Shi seemed to know everything about tea—its history, preparation,

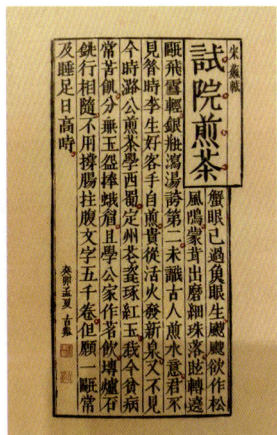

《试院煎茶》(苏轼)

① "胜若登仙不可攀，输同降将无穷耻"就出自范仲淹的《和章岷从事斗茶歌》。这首"斗茶歌"生动地描绘了宋代斗茶的情景和乐趣。

② 陆游出身茶乡、当过茶官，晚年亦归隐茶乡。其一生创作了200多首涉及茶的诗词，为历代诗人之冠。如名句"矮纸斜行闲作草，晴窗细乳戏分茶"，表明了写草书、玩分茶的闲适之情。

appreciation and medicinal functions. Besides, he even planted tea personally. Su wrote many inspiring and insightful poems and proses. In his prose *The Biography of Ye Jia*, he portrayed the life story of a man named Ye Jia, who was actually the portrait of the author himself as well as tea. "Prime tea is like sweet maiden", "Good brewing water needs to be boiled with burning flame", "No elixir of immortality can be compared to Lu Tong's seven bowls of tea", "Wine-drowsy when the road is long, I yearn for bed; Throat parched when the sun is high, I long for tea"[①]. With these sentences, Su Shi shared with us the charm of tea, the criterion of good brewing water, the health properties of tea, etc.

（8）改革者——朱元璋 / A Reformer in the Tea History— Emperor Zhu Yuanzhang

明代开国皇帝朱元璋（1328—1398）曾诏令"罢造龙团，惟采芽茶以进"，即将茶叶做成散茶进贡，废止了需要复杂制作工艺的饼茶。朱元璋出身贫寒，历经民间疾苦，此诏意为化繁为简，去奢靡之风，减轻百姓负担。朝廷风尚必然引领社会风尚，喝散茶的习俗至此成为主流。"废团改散"从根本上转变了中国的制茶工艺和饮茶习俗，顺应了大众的消费需求。至此，用沸水冲泡散茶成为流传至今的最常见的一种饮茶方式。

As the founding emperor of the Ming Dynasty, Zhu Yuanzhang ordered that "loose leaf tea be offered as tribute". Born into a poor family, Zhu Yuanzhang knew the life of common people. The making of tea cakes and preparation of whipped tea were both time-consuming and money-

---

① 该句为许渊冲先生所译。

consuming. This imperial edict meant to make tea drinking a simple and convenient practice. Catering to the demanding of ordinary people, loose leaf tea was accepted by all classes of society. Since then, brewing loose leaf tea with boiling water has become the most common way of tea consumption.

（9）高调茶人——乾隆皇帝 / The high-profile Tea Lover—Emperor Qianlong

生活在 18 世纪的清朝乾隆皇帝（1711—1799）是历史上著名的爱茶皇帝，他不仅规制茶器、评定用水^①、营建茶舍^②、组织茶会，还创作了 200 多首茶诗。乾隆皇帝在位期间举办宫廷茶宴——"三清茶宴"^③逾 40 次，推动了宫廷饮茶的规模和礼俗，引领当时茶文化的潮流与风尚。关于他的茶文轶事流传颇多，据说龙井茶、铁观音、君山银针等名茶都与他有关。乾隆皇帝在世 88 年，为历代皇帝之寿魁，这或许与他常年饮茶有些许关系。

Living in the 18th century, Emperor Qianlong (1711–1799) was a famous tea lover. As a versatile tea artist, he regulated tea utensils, tested brewing water, built tea rooms, and organized tea banquets. Besides, he wrote more than 200 poems about tea. During the reign of Emperor Qianlong, the "Sanqing Tea Banquet" was held more than 40 times, which promoted the popularity and innovation of tea drinking in the court. Today,

---

① 乾隆对水特别讲究，所用水包括玉泉山水、雪水和采集荷露所成的水。据说他特制一银斗，用以精量名泉之水，作为判断水质优劣的依据。

② 乾隆退位后，为继续品茶，专设"焙茶坞"。

③ 茶宴指每岁正月择吉日在重华宫举行以饮茶为主的宴席。始于清乾隆年间，乾隆、嘉庆、道光三朝共举行了 60 余次。茶宴所饮是以梅花瓣、佛手片、松仁烹制成的"三清茶"。但饮茶不过是茶宴的表面形式，实际内容是君臣赋诗联句，既联络君臣感情，又养生怡情。乾隆除为三清茶作诗外，还专门制作了三清茶碗。

there are many anecdotes about him who is often mentioned in stories of famous teas such as Xihu Longjing tea, Tieguanyin tea, Junshan Yinzhen tea. Qianlong lived for 88 years and was proud to be the oldest emperor in Chinese history. Perhaps it is justified to attribute his longevity in part to his love of tea.

重华宫茶宴（居中坐者为乾隆皇帝）

三清茶诗文盖碗

（10）文坛茶香——曹雪芹 / A Great Novelist and an Expert on Tea—Cao Xueqin

中国古典长篇小说四大名著之一《红楼梦》的作者曹雪芹（1715—1763），一生跌宕，披阅十载，完成百科全书式的巨作。以茶而言，《红楼梦》是我国历代文学作品中记述与描绘得最全的。全书120回，有112回提到茶，包括形形色色的名茶和饮茶方式、珍奇精美的古玩茶具，以及讲究的沏茶用水。此外，还有与茶相关的诗词（联句）十来首。《红楼梦》中的茶诗、茶文不仅数量多，且风格多样，既饱含生活气息，又极富艺术魅力，故而有人说："一部《红楼梦》，满纸茶叶香。"

茶香满红楼

A Dream of Red Mansions[①] is one of the Four Great Classical Novels of Chinese literatures. To some extent, the novel is based on the legendary life of its author, Cao Xueqin (1715–1763), who spent 10 years on this encyclopedic masterpiece. As far as tea is concerned, *A Dream of Red Mansions* reveals the most comprehensive picture of tea custom and culture. Among the total 120 chapters, 112 of them mention tea, including different teas, various tea drinking methods, exquisite antique tea sets and specially chosen brewing water. In addition, there are more than 10 poems (couplets) about tea. Because of vivid and comprehensive descriptions of tea, *A Dream of Red Mansions* is admired as "a novel full of aroma of tea".

---

① 《红楼梦》的英文书名，王际真译为 *Dream of the Red Chamber*；杨宪益和戴乃迭译为 *A Dream of Red Mansions*；英国汉学家霍克思翻译成 *The Story of the Stone*。本书采用 *A Dream of Red Mansions* 的译法。

（11）当代茶圣——吴觉农 / The Tea Sage of Modern China—Wu Juenong

出生于 19 世纪末的吴觉农（1897—1989）是一位农业科学家和经济学家，他在茶学理论、科研育人、产销贸易等领域均做出开创性贡献，被誉为当代中国茶业复兴和发展的第一人。即使在最动荡的岁月里，他亦全身心地投入中国茶业的自救与发展[①]。他推动了中国第一个高等院校茶学专业的建立[②]，创建了中国第一个全国性茶叶研究所[③]及新中国第一家全国性茶叶公司[④]，其所著《茶经述评》是研究陆羽《茶经》最权威的著作，被誉为"当代茶经"。因其为中国茶业发展做出的杰出贡献，吴觉农被誉为"当代茶圣"。

Born at the end of the 19th century, Wu Juenong (1897–1989) is an agricultural scientist and economist. He made pioneering contributions in many aspects of tea industry including scientific research, education, processing techniques, marketing and foreign trade. Even in the tumultuous years, Wu was firmly committed to the reform and development of Chinese tea industry. He promoted to set up the first tea major of higher education in China, established China's first tea research institute, and set up the first national tea company. His masterpiece *Review of the Classics on Tea* is so authoritative that it is reputed as "the Modern Classics on Tea". Because of

---

① 1922 年，吴觉农发表《茶树原产地考》，论证中国是茶树原产地。

② 1940 年，复旦大学创设茶业组和茶业专修科，吴觉农兼任系主任。1952 年，复旦大学茶业专修科迁至芜湖，并入安徽大学农学院，后农学院又迁至合肥独立建院（今安徽农业大学），茶专改为四年制的茶业系。

③ 1941 年，在吴觉农的推动下，中国第一个全国性茶叶研究所落户在福建省崇安县（1989 年崇安县撤县改市，现为武夷山市）。

④ 1949 年 12 月，中国茶叶公司（1985 年改为"中国茶叶进出口公司"）成立，时任茶业部副部长的吴觉农兼任总经理，统管茶叶生产、收购及外销业务。

his outstanding contribution to the development of Chinese tea industry, Wu Juenong is horned as the "tea sage of modern China".

小贴士

1. 冈仓天心在 *The Book of Tea* 中，对七碗茶诗的英译如下：The first cup moistens my lips and throat; the second cup breaks my loneliness; the third cup searches my barren entrails but to find therein some five thousand volumes of odd ideographs; the fourth cup raises a slight perspiration—all the wrong of life passes away through my pores; at the fifth cup I am purified; the sixth cup calls me to the realms of the immortals; the seventh cup—ah, but I could take no more! I only feel the breath of cool wind that rises in my sleeves.

2. 唐代元稹所创作的《一七令·茶》是一首宝塔诗。该诗描绘了茶的形态、功用和人们对它的喜爱之情，同时该诗还具有别致的形式美。

茶。

香叶，嫩芽。

慕诗客，爱僧家。

碾雕白玉，罗织红纱。

铫煎黄蕊色，碗转曲尘花。

夜后邀陪明月，晨前独对朝霞。

洗尽古今人不倦，将知醉后岂堪夸。

02 茶之类
Classification of Tea

2.1 茶叶分类方法 / Criteria for Classification

茶叶是以山茶科山茶属茶树上采摘的新梢为原料，采用特定加工工艺制作，供人们饮用或食用的产品。茶树广泛种植于亚洲、非洲和南美洲。中国、印度、肯尼亚、斯里兰卡、土耳其、印度尼西亚等是主要的茶叶生产国。中国是世界最大的茶叶生产国，在全国 34 个省级行政区中，产茶的有 20 多个[①]。茶叶种植与地理位置和自然环境密切相关，中国的茶区主要分布在北纬 18°~37°、东经 94°~122° 的范围内，分为江南、江北、华南、西南四大茶区。

Tea is produced from the leaves of the plant Camellia sinensis (L.) O.

---

① 据统计，2024 年，茶园面积超过 500 万亩（约合 33 万公顷）的省份有 4 个，分别是云南省、贵州省、四川省和湖北省。

Kuntze[1]. The species is now widely cultivated in Asia, Africa and South
America. Major producers of tea include China, India, Kenya, Sri Lanka,
Türkiye and Indonesia, among them China is the largest tea producer. Over
20 out of 34 provincial administrative regions in China produce tea. They are
divided into four tea-producing regions: Jiangnan (the South of the Yangtze
River), Jiangbei (the North of the Yangtze River), South China and Southwest

灌木型茶树（图虫　供图）

① 茶树的学名首先由瑞典植物学家林奈于 1753 年命名为 Thea sinensis；几经修改后，
1950 年我国植物学家钱崇澍确定茶树学名为［Camellia sinensis (L.) O. Kuntze］，并沿用
至今。

云南乔木型茶树（静观　供图）

China, each with different climates and geographical features. The cultivation of tea is closely associated with geographical location and natural environment, resulting in a distribution range between 18°–37° N and 94°–122° E.

　　茶树品种繁多，其外形和内含物质不尽相同，适合产制的茶类也不同[①]。此外，中国茶文化博大精深、源远流长，在漫长的历史发展过程中，历代茶人研制、开发了丰富多彩、各具特色的茶类。直至今天，中国已有 2000 多种茶品。

　　There are different types of tea plants, each with its own character and potential for unique cup quality. In addition, since tea has been processed and consumed by Chinese for thousands of years, there are a variety of

---

　　① 　如按树形分类，茶树有乔木、小乔木和灌木型三种；如按叶片大小分类，可分为特大叶类、大叶类、中叶类和小叶类；如按发芽迟早，可分为早芽种、中芽种和迟芽种。

processing techniques. All these contribute to the profound Chinese tea: tea with different colors, shapes, aroma and flavor. Presently, over 2,000 tea varieties are found in China.

目前茶有多种分类方法，如按茶叶发酵程度划分，分为全发酵茶、半发酵茶与不发酵茶；如按采摘季节划分，可分为春茶、夏茶和秋茶；还有按产地、销路、品质等分类。在诸多分类方法中，最常见的分类是基于茶叶的制作工艺，因为茶叶不同的颜色、形状、香气和滋味主要由制作工艺决定。据此，茶首先分为基本茶和再加工茶两大类。

Presently, there are different ways for the classification of tea. According to the degree of fermentation or oxidation, tea is mainly divided into three types: fermented tea, semi-fermented tea and non-fermented tea; according to the picking season, tea can be divided into spring tea, summer tea and autumn tea. Also tea can be classified according to the place of origin, sales channels and quality etc. Among them, the most popular one is based on the way tea is made, because it is the processing tea leaves undergo that determines its color, shape, aroma and flavor. Accordingly, tea is first divided into two categories—basic tea and reprocessed tea.

小叶种茶（静观　供图）

大叶种茶（陈燕　供图）

2.2 六大基本茶类 / Six Basic Types of Tea

基本茶类可进一步分为绿茶、白茶、黄茶、青茶（乌龙茶[①]、红茶和黑茶，主要是根据冲泡后汤的颜色而命名。六大茶类的分类标准主要是加工工艺和品质特征[②]。各类茶的茶青相差不大，不同的制作工艺造就了不同的茶类。中国是世界上唯一一个生产六大茶类的国家。

Based on the processing techniques and features of products, basic tea can be divided into six types—green tea, white tea, yellow tea, oolong tea, black tea and dark tea. Generally, teas all start off pretty much the same. It's the magic that happens during the tea production process that transforms them. China is the only country which produces the six types of tea.

多种基本茶类

① 青茶更为人所知的名字是乌龙茶，下文皆用乌龙茶之名，英文皆用 oolong tea。
② 《茶叶分类》国际标准（ISO 20715:2023）2023 年发布。该标准的发布标志着中国六大基本茶类分类体系成为国际通用标准。

（1）绿茶 / Green Tea

绿茶属于不发酵茶，基本工艺流程包括杀青、揉捻、干燥。杀青[①]是绿茶制作的核心工序，是形成和提高绿茶品质的关键性技术措施。在揉捻阶段，根据不同产品特点，茶叶被制成不同的形状[②]。结合工艺特点，绿茶通常分为炒青绿茶、烘青绿茶、晒青绿茶和蒸青绿茶[③]。绿茶的制作工艺决定其较多地保留了鲜叶内的天然物质，冲泡后茶叶仍然呈绿色，茶汤清澈。

Green tea is non-fermented tea. The processing techniques include fixation, rolling and drying. The process of fixation is crucial to the quality of green tea. According to different ways of fixation and drying, green tea is subdivided into pan-fried green tea, basket-fried green tea, sun-dried green tea and steamed green tea. While in the rolling process, tea leaves are rolled into different shapes. To make green tea, the fermentation or oxidation process is completely left out, which gives green tea light, fresh flavor and delicate color.

绿茶是我国主要茶类之一，历史悠久、茶区广、产量高、品质好。中国是全球最大的绿茶生产、消费和出口国。我国名优绿茶很多，包括西湖龙井茶、洞庭（山）碧螺春茶、黄山毛峰等。高档名优绿茶对原料要求极高，如特级西湖龙井茶、洞庭（山）碧螺春茶的鲜叶非常细嫩，制作 500 克干茶需数万个嫩芽头[④]。

---

① 杀青的主要作用是迅速破坏鲜叶中酶的活性，防止多酚类物质大量氧化，保有绿茶特有的色香味。

② 如针形的雨花茶、圆珠形的珠茶、扁形的龙井、卷曲形的碧螺春等。

③ 杀青分为加热杀青和热蒸汽杀青，据此绿茶分为炒青和蒸青；干燥分为炒干、烘干和晒干，据此绿茶可分为烘青和晒青。

④ 很多名茶是手工采摘鲜叶、人工炒制，以保证茶的品质，故价格也比较高。

西湖龙井（绿茶）

Green tea has a long history and it is mainly produced in China[1]. As the world's largest green tea producer, consumer and exporter, China abounds with numerous famous green teas such as Xihu Longjing tea, Dongting Biluochun tea, Huangshan Maofeng tea, etc. The quality of fresh tea leaves plays a key role in the quality of green tea. For example, top Xihu Longjing tea and Dongting Biluochun tea are processed from the finest tender fresh tea leaves. Tens of thousands of tender buds are needed to make 500 grams of them.

（2）白茶 / White Tea

白茶是制作工序最少的微发酵茶，基本工艺流程包括萎凋、晒干或烘干。萎凋是白茶加工的关键工艺，指通过晾晒等方式，使鲜叶呈现萎蔫状态，其主要目的是降低鲜叶含水量。从某种程度上说，萎凋也是一种自然氧化。

White tea is a lightly fermented tea with the simplest processing procedures including only prolonged withering and drying. When fresh leaves arrive at the factory, they are either left to wither in the sun or indoors. The function of this process is to reduce water content. To some extent, withering is a natural way of oxidation.

---

[1]　Steamed green tea is mainly produced and consumed in Japan.

白茶干茶表面覆盖一层白色茸毛，因此得名。根据茶树品种和原料的不同，白茶分为白毫银针、白牡丹、贡眉、寿眉四种。白茶的品质体现在特有的香气和滋味上，新白茶以清鲜、醇爽、甘甜、毫香足为佳，老白茶以醇厚、回甘显为佳。白茶追求后期的转化，其口感、功效、价格都随着时间的增长而变化。因白茶加工工序最少，故被视为最健康的茶类，深受顾客喜爱；近年来老白茶的保健效果也成为市场热点。白茶主要产于福建省福鼎市、政和县等地，其中福鼎市有"中国白茶发源地"之称①。

白毫银针（白茶）（静观　供图）

Tender tea leaves are covered with a layer of white fluff, hence the name "white tea". According to the quality of tea leaves, white tea is divided into Baihao Yinzhen tea (Silver Needle with White Hair tea), Bai Mudan tea (White Peony tea), Gong Mei tea (Tribute Eyebrow tea) and Shou Mei tea (Longevity Eyebrow tea). White tea is known for its unique aroma and taste. The newly made white tea tastes fresh, sweet and fragrant while the aged white tea is mellow with rich after-taste. If it is stored properly, white tea can be kept for a long time and its taste and function change or improve with time, so does the price. Since

---

① 由于喜爱白茶的人越来越多，安徽、浙江等省份，印度、斯里兰卡、肯尼亚等国家也开始生产白茶。

white tea undergoes minimal processing, it is considered the healthiest tea type. In recent years, the medicinal value of aged white tea has been recognized by the public. White tea is mainly produced in Fuding and Zhenghe in Fujian province, and the former is known as the "birthplace of white tea".

（3）黄茶 / Yellow Tea

　　黄茶属于轻发酵茶，基本工艺流程与绿茶相似，增加了特有的闷黄工序，这是形成黄茶"黄汤黄叶"特点的关键。因干茶和茶汤都有独特的黄色，故名黄茶。黄茶香气独特，口感清香醇厚。黄茶按鲜叶老嫩、芽叶大小又分为黄芽茶、黄小茶和黄大茶。湖南的君山银针、四川的蒙顶黄芽是黄芽茶的代表，也是最知名的黄茶。黄茶以内销为主，是六大茶类中最小的一类，也是中国特有的茶类。

霍山黄大茶（黄茶）（静观　供图）

Yellow tea is a kind of lightly fermented tea. Its processing technique is similar to that of green tea, except for a special process called yellowing. The process of yellowing gives yellow tea unique taste and color. Both dry tea leaves and infusion of yellow tea are yellow, hence the name. Yellow tea is subdivided into bud yellow tea, small-leaf yellow tea and large-leaf yellow tea according to the plucking standards and tenderness of the fresh leaves. Junshan Yinzhen tea (Junshan Silver Needle tea) from Hunan province and Mengding Huangya tea (Mengding Yellow Bud tea) from Sichuan province are the rarest and most celebrated creations of our country. Yellow tea is the smallest group in basic teas, which is only produced in China for domestic consumption.

（4）乌龙茶 / Oolong Tea

乌龙茶是一种半发酵茶[①]，基本工艺流程包括萎凋、做青、炒青、揉捻（或包揉）、烘焙，做青是乌龙茶特有的关键工序，由摇青和晾（凉）青交替进行。乌龙茶融绿茶的清香与红茶的醇厚为一体，干茶呈特有的绿褐色或褐色。冲泡后，茶叶呈绿叶红镶边，而茶汤则呈橙黄色或黄色，滋味浓郁、韵味独特。乌龙茶独特的品质特征是特定的生态环境、茶树品种和采制技术综合作用的结果。

Oolong tea sits somewhere between the green and black teas in that it is partly fermented. The making techniques includes withering, Zuoqing, panning, rolling and firing. The process of Zuoqing is crucial to the quality of oolong tea, consisting of shaking and cooling of green leaves, which gives oolong tea characteristic appearance and taste. The color of oolong tea leaves is distinctively greenish brown or coppery red. After being steeped in boiling water, the tea leaves are showing green in the center with red edge, while the liquor[②] looks yellow-red or yellow. The unique charm of oolong tea is the result of the specific ecological environment, different varieties of tea plants and processing techniques.

---

① 按发酵程度，乌龙茶可分为"轻发酵乌龙茶""中发酵乌龙茶"和"重发酵乌龙茶"。
② 茶汤可译为 infusion、tea liquid、liquor 等，文中较多采用 liquor (referring to the water after brewing with tea leaves)。

乌龙茶源于福建，目前主要产于福建、广东和台湾三省，因产地及品种品质上的差异，乌龙茶分为闽北乌龙茶、闽南乌龙茶、广东乌龙茶和台湾乌龙茶。四大产区的代表茶品分别是武夷岩茶[①]、安溪铁观音、凤凰单丛[②]及台湾高山乌龙茶和文山包种。

Originated in Fujian province, oolong tea is mainly produced in Fujian, Guangdong and Taiwan provinces. There are many varieties and the popular ones include Wuyi Yan Cha (Wuyi Rock Tea), Anxi Tieguanyin tea, Fenghuang Dancong tea (Phoenix Single Bush tea), Taiwan Mountain oolong tea and Wenshan Baozhong (Pou chong) tea.

台湾乌龙茶（乌龙茶）（静观　供图）

（5）红茶 / Black Tea

红茶是一种全发酵茶，生产工艺包括传统工艺及 CTC 工艺。在中国，红茶加工较多采用传统工艺，其基本流程包括萎凋、揉捻（切）、

---

① 武夷岩茶中的名品包括有"岩茶之王"美誉的大红袍、水金龟、铁罗汉、白鸡冠、肉桂、水仙等。

② "凤凰"原是指茶树生长在广东省潮安区凤凰山，"单丛"是指单株采摘、单株制茶、单株销售。凤凰单丛的特点主要体现在其香气，以花香和果香为主，故常用茶的香气分类命名，如蜜兰香、芝兰香等。

发酵、干燥 ①。CTC 是碾碎（crush）、撕裂（tear）、卷起（curl）三个英文词的缩写，该工艺是一种以快速、低成本、大量生产红茶为目的的制法，出现于 20 世纪 30 年代，可在相同包装前提下，增加茶叶的实际容量 ②。CTC 工艺的特点在于，鲜叶萎凋后送入 CTC 机器，经过碾压、切碎、揉卷成极细小的圆粒状，便于制作茶包，随后的发酵和干燥流程与传统工艺相似。无论采用哪种工艺，发酵都是形成红茶色、香、味品质特征的关键工序。

Black tea is a fully fermented tea. There are two ways of making black tea, the orthodox method and the CTC method[3]. In the orthodox method which is mainly used in China, the processing techniques include withering, rolling, oxidation and drying. CTC which means Crush, Tear and Curl, is an efficient way of mass production of black tea at low cost. It was invented in 1930s to increase the weight of tea that could be packed into a sack or chest. In this process, the leaves are first withered and then they are put through a series of rollers, covered in hundreds of small, sharp teeth. These teeth cut, tear and curl the leaves, producing tiny granules which are perfect for tea bags. After this, they go through the same oxidation and drying processes as the orthodox method. In both methods, oxidation contributes greatly to the distinctive color, taste and strength of black tea.

中国红茶产地广、种类多，以生产工艺和产品特性进行分类，红茶可分为小种红茶、工夫红茶和红碎茶。产于福建武夷山的小种红

① 文中所列加工流程为传统工艺；工夫红茶多一步精制加工工艺，而小种红茶的加工工艺突出熏松烟。
② 根据实际冲泡比较，1千克CTC茶叶约可冲泡四百杯红茶，1千克传统制法茶叶能冲泡约200 杯。
③ Most of the black tea in India and Kenya is manufactured in the CTC process, while black tea in China is mainly processed by orthodox rollers.

茶——正山小种是世界红茶的鼻祖。小种红茶特有的松烟香源于在干燥阶段采用松柴烟熏的独特工艺。小种红茶根据产地、加工和品质的不同，可分为正山小种和烟小种两种产品。①

Black tea is produced in many places in China, and according to processing technique and tea features it can be subdivided into Souchong black tea, Congou black tea and Broken black tea②. Souchong black tea includes Lapsang Souchong and smoky Souchong. Made in the Wuyi Mountain, Fujian province, Lapsang Souchong is the world's first black tea. There isn't a scent or flavor more distinctive than that of Souchong. It's all down to the pinewood smoke that permeates the tea leaves when they're being dried. According to differences in origin, processing methods, and quality characteristics, Xiaozhong black tea is classified into two distinct varieties: Lapsang Souchong (Zhengshan Xiaozhong) and Smoked Souchong (Yan Xiaozhong).

工夫红茶是条形茶，通常按产地命名，如祁门工夫、滇红工夫、白琳工夫等，其中有"群芳最"之美誉的祁门工夫是世界三大高香红茶之一③。

China is known for a variety of Congou black teas which are usually named according to their origins, such as Keemun (Qimen) Congou tea,

---

① 根据国家标准《红茶 第3部分：小种红茶》（GB/T 13738.3—2012），正山小种是指产于武夷山市星村镇桐木村及武夷山自然保护区域内的茶树鲜叶，用当地传统工艺制作，独具似桂圆干香味及松烟香的红茶产品；而烟小种是指产于武夷山自然保护区域外的茶树鲜叶，以工夫红茶的加工工艺制作，最后经松烟熏制而成，具松烟香味的红茶产品。

② Souchong black tea, Congou black tea and Broken black tea 的翻译分别采用国标红茶第1部分：红碎茶（GB/T 13738.1—2017）；第2部分：工夫红茶（GB/T 13738.2—2017）；第3部分：小种红茶（GB/T 13738.3—2012）的翻译。

③ 印度大吉岭红茶、斯里兰卡乌瓦红茶、中国祁门红茶被誉为世界三大高香红茶。

Dianhong Congou tea and Bailin Congou tea etc. Among them, Keemun tea, with the reputation of "the most fragrant tea", is one of the top three famous black teas in the world (the other two being Darjeeling tea from India and Uva black tea from Sri Lanka).

祁门红茶（红茶）（静观　供图）

　　红碎茶[①]是在加工时将鲜叶切碎而制成的（碎、片、末状）颗粒型红茶。红碎茶冲泡时，茶汁浸出快、浸出量大，为方便冲泡，通常被加工成茶包。茶包虽然会使茶叶的品质有所下降，但消费者却愿意选择这种便利性。作为目前世界上消费量最大的茶叶品类，红茶味道甜美醇厚，既可清饮，也可加入牛奶、糖等调饮。

Broken black tea is made by chopping tea leaves into granules and it is usually made into tea bags which provide a convenient and popular way of tea drinking. Tasting sweet and mellow, black tea is the most popular tea in the world. Some people prefer plain tea without adding anything, while others love to add various ingredients like milk and sugar into the tea.

---

① 红碎茶依据外形，分为叶茶、碎茶、片茶及末茶。

需要说明的是，"black tea"① 所指在中英语言文化中有一定的差异。大多数红茶的干茶颜色偏黑，但冲泡后，红茶的茶汤和叶底呈红色，故重其汤色的中国人称其为红茶，而重其干茶颜色的欧美人以 "black tea" 呼之。故英语中的 "black tea" 实际指的是中国的红茶。为了与国际市场保持一致，我们接受 "black tea" 的用法，相应的，我们用 "dark tea" 指代中国的黑茶。"black tea" 的指代歧义，从另一个角度反映了中国茶的多样性。

It is noteworthy that the meaning of "black tea" is somewhat ambiguous. "Black tea" equals Chinese "hong cha" or Chinese "red tea". This is because most black teas are dark brown or black in term of the color of dry tea leaves. While in turns of the color of the infused liquid, it is somewhat red or reddish and that's why the Chinese call it "hong cha" or "red tea". In order to be consistent with the international usage, Chinese use "black tea" to refer to Chinese "hong cha" or "red tea" and Chinese "Hei cha", literally Chinese black tea, is therefore called "dark tea". Interestingly, the ambiguity of "black tea" reflects the diversity of Chinese teas.

（6）黑茶 / Dark Tea

黑茶② 是后发酵茶，基本工艺流程包括杀青、揉捻、渥堆③、干燥等。渥堆是黑茶品质形成的重要工序，也是形成黑茶独有的醇厚顺滑口感的关键。黑茶是中国特有茶类，一般原料较粗老。因其在制作过

---

① 最早进入欧洲的小种红茶干茶色泽乌黑油润，当地人称为"乌茶"，black tea 正是产地方言"乌茶"的英译。

② 黑茶历史悠久，早期的蒸青团饼绿茶由于长时间的烘焙干燥和非完全密封运输贮存，在湿热氧化作用下由绿色变褐色，成为黑茶的雏形。

③ 渥堆指在一定的温、湿度条件下，通过茶叶堆积，促使其内含物质缓慢变化的过程。

程中往往堆积发酵时间较长，因而叶色油黑或黑褐，故称黑茶。传统上，黑茶按生产地域及产品特性主要分为湖南黑茶、湖北黑茶、四川黑茶、广西黑茶和云南普洱熟茶等；而最新的《茶叶分类》国际标准（ISO 20715:2023）则将黑茶分为普洱熟茶和其他黑茶。适于收藏、可以长期存储[①]是黑茶的一大特色。在存储得当的前提下，黑茶以陈为贵，这一特征与白茶相似，即储存年份越久，茶味越醇厚香浓。

多种黑茶　　　　　　　　广西六堡茶（黑茶）（静观　供图）

Dark tea is a kind of post-fermented tea. The processing techniques include fixation, rolling, pile-fermentation and drying. The process of pile-fermentation is crucial to the quality of dark tea and contributes greatly to its distinctive characteristics. Dark tea is mainly made in China. Generally, the fresh leaves used to make dark tea are more mature than those for making green tea or black tea. Due to the process of pile-fermentation, tea leaves turn dark and black, hence its Chinese name "Hei cha", literally "black tea". According to the production region and making techniques, dark tea is traditionally subdivided into Hunan dark tea, Hubei dark tea, Sichuan

---

① 在符合包装、运输、储存要求的前提下，黑茶可长期保存。但并非所有黑茶都能越陈越香，"原料是基础，工艺是关键，仓储是升华"。

dark tea, Guangxi dark tea and Yunnan ripened Pu'er tea. While according to Classification of Tea Types (ISO 20715:2023), dark tea is divided into ripened Pu'er tea and other dark teas. Dark tea can be stored for a long time due to its special processing techniques. Just like white tea, the flavor of dark tea mellows with time if stored well.

2.3 再加工茶类 / Reprocessed Tea

再加工茶是以基本茶类的茶叶为原料经过不同的加工工艺而形成的茶叶类别，包括花茶、紧压茶、袋泡茶、粉茶等几类①。花茶和紧压茶是再加工茶的两个主要代表。

口味丰富的再加工茶

① 国际知名品牌常常将茶分为常规茶和特色茶。特色茶也可以理解为加味茶，既可加草本味，也可加果味等。世界范围内最知名的加味茶主要有茉莉花茶和格雷伯爵红茶，前者颇受中国消费者喜爱，后者以微酸的佛手柑平衡红茶的醇厚，是国际市场最受欢迎的加味茶。(It is generally believed that artificially flavored teas such as apple tea belong to specialty teas and the non-flavored ones the regular teas.)

自制梅香安吉白茶（马娜　供图）

Reprocessed tea refers to any of the regular or basic tea being further processed. It mainly includes scented tea, compressed tea, bagged tea, and powdered tea. Among them, scented tea and compressed tea are the two representatives.

花茶①是利用茶叶的吸附性，由茶叶和花卉配制而成。茶坯通常选用绿茶，亦会用红茶或乌龙茶等茶类；花通常选用茉莉花、桂花、玫瑰、梅花、菊花、莲花等。根据茶吸收的香味，花茶可分为茉莉花茶、

① 花茶，又名窨花茶、熏花茶或香片。花茶源于宋、始于明而成于清。

玫瑰花茶等。花茶的质量取决于茶的质量、鲜花的品质和窨制工艺[①]。好的花茶既有茶香，又有花香，味道甜美，令人愉悦。

Scented tea is a mixture of tea and flowers. Usually we use green tea, black tea or oolong tea to make scented tea. In terms of flowers, we often use jasmine, sweet osmanthus, rose, plum flowers, chrysanthemum and lotus flowers, etc., so scented tea is usually named after the flower used. High quality scented tea requires good tea and flowers. The right proportion of tea to flowers will enhance the taste of the scented tea. Sweet, pleasant and delightful, scented tea has both the fragrance of tea and aroma of flowers.

我国北方人，尤其是北京人特别喜欢喝花茶，颇为钟爱茉莉花茶[②]。茉莉花茶[③]是花茶中产销量最多的品种，以福建福州[④]、宁德和江苏苏州所产的品质最好。茉莉花茶加工非常耗时。春天茶坯做好后保鲜存放，到了三伏天茉莉花开的时候，将茶坯及正在吐香的茉莉花层层堆放，使茶叶吸收花香。

People in northern China are especially fond of jasmine tea. As the most popular scented tea in China, Jasmine tea made in Suzhou of Jiangsu province and Fuzhou and Ningde of Fujian province are of the best quality. Making jasmine tea is time consuming. The leaves are harvested and processed in

---

① 加工时，将茶坯和鲜花充分均匀混合，使茶叶吸收花香；待鲜花的香气被吸尽后，挑出花卉，再换新的鲜花按上法窨制。花茶香气的高低，取决于所用鲜花的质量、数量和窨制的次数，窨制次数越多，香气越高。

② 国家级非遗名录中，花茶制作技艺有三项入围，而"张一元茉莉花茶制作技艺""吴裕泰茉莉花茶制作技艺"均为北京所申报，可见北京是国内茉莉花茶制作和消费的高地。

③ 对上海豫园地区茶相关店铺及上海入境游导游群体调研显示，入境游客最青睐的茶品为茉莉龙珠；其他受外国友人喜爱的包括工艺花茶、人参乌龙茶等。

④ 2014年，"福州茉莉花茶窨制工艺"入选国家级非遗名录。

spring, then kept fresh till the dog days, when the young jasmine buds are picked and layered over the tea to release their essential oils.

茉莉花茶

紧压茶是将散茶作为原料，通过模具压制而成，因紧压茶通常呈砖状或饼状，故常被称为砖茶或饼茶。根据所用原料^①茶类不同，可分为黑茶紧压茶、白茶紧压茶、乌龙茶紧压茶等，其中以黑毛茶为原料制成的黑茶紧压茶最为常见，如云南七子饼茶、湖南千两茶、广西六堡茶^②等。

Compressed tea is usually made by compressing primary tea leaves into solid cakes and bricks; hence the name tea brick or tea cake. According to the primary tea used, it can be divided into compressed dark tea, compressed white tea and compressed oolong tea, etc. Among them, compressed dark tea is the most widely consumed type, and Pu'er Qizi tea cake from Yunnan, Qianliang tea from Hunan, and Liubao tea from Guangxi are quite familiar to customers.

---

① 紧压茶的原料多数比较粗老，干茶色泽黑褐，汤色橙黄或橙红。

② 七子饼茶又称圆茶，茶叶紧压成圆饼状，七饼装为一筒，故名七子饼茶。七子饼茶以晒青毛茶和普洱散茶为原料进行压制，有生饼和熟饼之分。千两茶也叫花卷茶，是柱形紧压茶，以每卷茶叶净含量合老秤1000两而得名。六堡茶原产于广西苍梧县六堡镇，因此得名。六堡茶是篓装紧压茶，具有"红、浓、醇、陈"的特点。

紧压茶

我国古代就有紧压茶，如唐代的蒸青团饼茶和宋代的龙凤团茶。紧压茶是传统的边销茶，内销西藏、新疆、内蒙古、甘肃等地，外销俄罗斯、蒙古等国。

Compressed tea has a long history; it is believed that the steamed tea cake in the Tang Dynasty and Longfeng (dragon and phoenix) tea cake in the Song Dynasty were ancestors of compressed tea. Compressed tea is convenient for transportation and storage, thus a favorite for people in remote regions such as Xizang, Xinjiang and Inner Mongolia. It is also welcomed by people in neighboring countries such as Russia and Mongolia.

第二部分
茶韵风雅
A Sip of Tea,
A Taste of Life

03 茶之泡
Tea Brewing

　　在数千年的饮茶历史中，饮茶已经成为人们日常生活的一部分。对于中国人而言，泡茶是一门艺术，需要高超的手法呈现茶的魅力。泡茶也是一门科学，其本质上就是将茶叶内含成分充分浸出至茶汤中的过程。茶与水的质量、茶具的选择、冲泡的方法等直接影响茶叶内含成分的浸出量，影响茶汤的色、香、味及其保健效果。

　　Tea has been an important part of everyday life of Chinese people for thousands of years. For Chinese, brewing tea is an art and a skill as well which helps bring out the essence of tea. Among many things that may affect tea brewing, tea, water, tea ware and brewing skills are the four most important elements. They are directly related to the leaching of constituents in tea, which affects the color, aroma, taste, and health benefits of the liquor.

3.1 冲泡的基本原则 / Basic Skills of Tea Brewing

（1）泡茶用水 / Brewing Water

"水为茶之母"，中国人非常重视泡茶用水。明代张大复认为："八分之茶，遇十分之水，茶亦十分矣；八分之水，试十分之茶，茶只八分耳。"泡茶用水究竟以何为好？自古以来，人们就在不断探寻这个问题的答案[①]。历史上有不少鉴水的逸事佳话，如陆羽、苏轼、乾隆皇帝品泉择水的故事，也有不少关于试水烹茶的诗文流传至今。如唐代的茶圣陆羽认为："其水，用山水上，江水中，井水下。"虽然古人对水的选择有不同的看法，但他们普遍认为泉水、雨水和雪水是上好的泡茶用水。

Half of the art of making tea lies in getting good water. For Chinese people "water is mother of tea"[②]. Such is the importance of water to brewing tea that a scholar named Zhang Dafu in the Qing Dynasty once said "Common tea brewed with good water, the liquor may reach its perfection; while good tea brewed with common water, the liquor is common." What are characteristics of good water? To this question, the Chinese have been in search of the answer since ancient times. There are many poetic and imaginary descriptions of how tea and water can make a perfect match. Among them, stories about Lu Yu, the tea sage, Su Shi, a famous poet and

---

[①] 宋徽宗赵佶在《大观茶论》中提出水以"清、轻、甘、洁"为美；苏东坡则认为烹茶用水，以"活"为贵，"活水还须活火烹，自临钓石取深情"；而唐庚在《斗茶记》中也指出"水不问江井，要之贵活"。

[②] 这句话的翻译化用了英国诗人 William Wordsworth 的名句"The child is father of the Man"。

Qianlong, a tea-loving Emperor, are quite popular. Back to the 8th century, Lu Yu dwelled on the choice of water by saying "the mountain spring is the best, the river water comes next and well water comes third in order of excellence". Generally, spring water, rainwater and melted snow were favored in tea brewing in ancient times.

　　泡茶用水的选择可大致归纳为水质和水味两大要素。从水质的角度来看，水要活、清和轻；从水味的角度来看，水要甘而冽。简而言之，好的泡茶用水须是活水，须洁净甘甜。

The characteristics of good water lies in two aspects: the quality and the taste. In terms of quality, water should be fresh, clear and light; in terms of taste, it should be sweet and refreshing. In short, good water must be fresh, sweet, clean and clear.

（清代）蒸馏水器

　　现代人在泡茶时，水的选择更多[①]且取用更为方便。生活在乡村的人，可以选择山泉水[②]等优质自然水源；生活在都市的人，可以选用瓶装矿泉水、纯净水或蒸馏水，也可以选用煮沸的自来水。但自来水中

---

　　① 　一般泡茶用水要求清洁、无异臭和异味，水的硬度不超过8.5度（根据水中所含钙、镁离子的多少，水分为硬水和软水。饮茶用水以软水为好，用软水泡茶，茶汤明亮、滋味鲜爽），色度不超过15度，pH值在6.5左右，不含肉眼所能看到的悬浮微粒，不含有腐败的有机物和有害的微生物，浑浊度不超过5度，其他矿物质元素含量均要符合我国《生活饮用水卫生标准》（GB 5749-2022）的要求。

　　② 　镇江中冷泉、无锡惠山泉、苏州观音泉、杭州虎跑泉和济南趵突泉，号称中国五大名泉。

趵突泉水大碗茶（图虫　供图）

漂白粉的气味较重，会影响茶的香气和口感。为改善口感，我们可将自来水静置于干净容器中 24 小时，让氯气挥发，如此漂白粉的异味可基本消散，但因久置，水的含氧量降低，也会影响口感①。安装家庭净水器，接取新鲜的水煮沸泡茶，是家庭泡茶用水的上选。

Nowadays, there are many kinds of brewing water available. Some people brew tea with water directly from mountain spring. For people living in cities, bottled mineral water, purified water or distilled water are preferred. Boiled tap water is a cheap and convenient choice. However, the smell of bleaching powder in tap water spoils the aroma and taste of tea. Keeping the water in a clean container for about 24 hours can volatilize the chlorine but much of the oxygen in the water will be lost at the same time. As we know, freshly boiled water contains more oxygen, which can bring out the essence of the tea. Therefore, the fairly convenient and economical way of getting brewing water is to install a water purifier at the kitchen.

（2）泡茶用具 / Tea Ware

"器为茶之父"，中国人认为茶具的选择对于冲泡也非常重要。合适的茶具不仅能凸显茶的美好，还能提升品饮体验。冲泡不同的茶叶

---

① 也有一些专家和知名品牌提出，泡茶需要新鲜的水，因其富含氧气，有利于激发茶汤的香气和滋味。

要选用不同材质、样式和颜色的茶具。首先，茶具的材质会影响滋味的呈现；其次，茶具的造型、色彩和整体风格要与选用的茶及茶人想要营造的意境相得益彰。

To emphasize the importance of tea ware to tea brewing, Chinese people say "Tea ware is father of tea". In other words, good tea ware complements tea; it can even enhance the tea drinking experience to the next level. The harmonious combination of the material, function, and color of tea ware is essential to brewing a good cup of tea. Firstly, tea ware with suitable material may improve the flavor of tea. Secondly, the shape, color, and overall style of the tea set need to be carefully matched with the tea, the tea brewer, and the desired artistic conception.

比如我们用散热快、不易产生熟汤气的透明玻璃茶具冲泡细嫩的绿茶，在欣赏茶叶沉浮的同时，感叹人生如茶，沉时坦然、浮时淡然。紫砂壶古朴典雅，利于激发滋味却不会破坏茶香，适合冲泡乌龙茶和黑茶。产于江西景德镇的瓷器，样式齐全、美观大方，清洗后不滞留气味，适于冲泡多种茶类。

银壶煮水泡茶（静观 供图）

玻璃杯冲泡绿茶（视觉中国　供图）

For example, one may appreciate unfurling and jumping of leaves in water when delicate green tea is brewed with transparent glass tea ware. Simple and elegant, the purple clay tea pot is often used for brewing robust teas such as oolong tea and dark tea. The porcelain tea ware made in Jingdezhen, Jiangxi province is well known for its beautiful texture and graceful design and it can be used to brew all types of teas.

（3）泡茶技艺 / Brewing Skills

　　有了优质的茶叶、甘甜清洌的用水、合适而精致的茶具，还需要高超的冲泡技艺，才能把茶的色、香、味充分地展现出来，给人以美的享受。冲泡技艺主要体现在对投茶量、冲泡水温和冲泡时间的掌握上。

A good cup of tea requires not only high-quality tea, sweet and clear water and proper tea ware, but also brewing skills. Only in this way, can we fully enjoy the beauty, flavor and aroma of tea. Brewing skills are mainly reflected in the mastery of "3T"—Tea dose, water Temperature and brewing Time.

烫杯

投茶量 / Tea Dose

要泡好茶，首先要掌握茶叶用量（或茶水比例）。茶叶用量并没有统一的标准，主要根据茶叶种类、茶具大小以及品茶人的饮用习惯而定。日常泡茶通常是凭个人的经验和习惯而定，一般来说茶与水的比例为1：50。但由于茶叶种类以及冲泡方式不同，茶叶用量也有差异，如用紫砂壶冲泡乌龙茶，茶叶用量明显多于用玻璃杯冲泡红茶或绿茶。

Brewing skills lie first in tea dose or the proportion of tea to water. There is no fixed standard for the proportion, which is mainly determined by the type of tea used, the size of the tea ware and personal drinking preference. The general proportion of tea to water is 1：50. However, the proportion of tea to water varies with different types of tea and brewing methods. For example, when brewing oolong tea in a purple clay pot, the amount of tea used is much more than that of black tea or green tea brewed with a glass.

置茶

冲泡水温 / Water Temperature

冲泡水温也会影响茶汤品质。冲泡水温因茶而异。首先，水温与茶的嫩度有关，冲泡细嫩的茶叶，水温较低，反之亦然，如冲泡明前龙井，用80℃左右的水即可。其次，冲泡水温还与茶叶种类或茶叶的发酵程度有关，一般而言茶叶的发酵程度越高，需要的水温也就越高。此外，冲泡水温也与茶叶中有效物质在水中的溶解度呈正相关，即水温越高，溶解度越大，茶汤越浓。

Traditionally, brewing water should be hot. But with different teas, the temperature can be quite different. Firstly, the tenderness of tea should be taken into consideration. For example, when brewing tender Longjing tea, the temperature can be as low as 80℃. Secondly, the more fermented the tea is, the hotter of brewing water is required. In addition, using water at higher temperature leads to greater solubility and stronger infusion.

在日常生活中，如果用滚开的水冲泡细嫩的绿茶，茶汤就会既苦又涩，若我们将水烧开后冷却5分钟再进行冲泡，便可品味绿茶的清雅；而红茶、乌龙茶和黑茶等发酵程度较高的茶，通常需用沸水冲泡以激发茶香。

For example, in daily life, far too often, green tea is over-brewed, becoming tannic and bitter. If we allow the boiling water to cool for about 5 minutes before pouring, we may taste the delicate and subtle flavor of green tea. When brewing black tea, oolong tea or dark tea, boiling water is often used to promote the fragrance and mellowness of the tea.

近年来，冷泡茶①逐渐被大众接受。所谓冷泡，即用常温或冷藏的矿泉水、纯净水等冲泡茶叶（通常用茶包），是一种颠覆传统的泡茶方法。

However, brewing water can also be cold. In recent years, cold water infusions have gained in popularity. It offers an easy and convenient way to appreciate the essence of tea. All one should do is to put the tea (usually the tea bag) into a bottle and then fill the bottle with cold water or water at room-temperaturc.

冲泡时间 / Brewing Time

冲泡时间也是影响茶汤浓度和滋味的因素之一。时间太短，茶汤寡淡；时间太长，茶汤苦涩。冲泡时间取决于茶叶种类、冲泡水温、投茶量和饮茶习惯。日常用玻璃杯冲泡红茶或绿茶，3分钟后饮用可获得最佳口感（也有人会加大投茶量，快速出汤，以获得更好的口感）。用小型紫砂壶冲泡乌龙茶，投茶量7~8克，第一泡1分钟左右就要出汤，从第二泡开始，每次冲泡时间都要比前一次延长15秒，这样前后茶汤的浓度会比较均匀，且因为出汤快，可多次冲泡。

---

① 有研究表明，茶叶冷泡后多酚类和咖啡碱的溶出率比热泡茶低，而茶氨酸和多糖类的溶出率要比传统热泡茶高，可减少茶汤苦涩味，增加鲜醇感；且降低茶汤中的咖啡因含量，可减缓其对胃的刺激性，因此敏感体质的人比较适合饮用。

Optimal brewing time also helps produce a beautiful and smooth cup. The brewing time depends on the type of tea, water temperature, the amount of tea and personal preference. Generally, short steeping time leads to weak and bland infusion; while too long steeping time results in strong and bitter infusion. For example, when brewing black tea or green tea with a glass, the recommended brewing time is about 3 minutes. When brewing oolong tea with a small purple clay pot, 7–8 grams of tea (a higher tea-to-water ratio) is proper. The first infusion should be poured out in a minute; an additional 15 seconds should be added for the following each brew; in this way, the infusion may taste almost the same.

总体来说，细嫩的茶叶比粗老的茶叶冲泡时间要短；松散的茶叶、碎末茶叶比紧压的茶叶、完整的茶叶冲泡时间要短；对于注重香气的茶叶，如乌龙茶和花茶，冲泡时间不宜长；而白茶加工时未经揉捻，茶汤较难浸出，冲泡时间则相应延长。通常茶叶可冲泡 3~7 次，这主要取决于投茶量、冲泡时间、茶叶种类和茶叶品质。

Generally speaking, the brewing time of fine and tender tea is shorter than that of coarse tea; the brewing time of loose tea or broken tea is shorter than that of compressed tea. For oolong tea and scented tea which are known for fragrance, the brewing time should be short while for white tea, long brewing time is needed due to its unique processing techniques. Usually, tea can be brewed for 3 to 7 times, which mainly depends on tea dose, the duration of each infusion, tea type and quality.

总之，在泡茶过程中，投茶量、冲泡水温和冲泡时间是最重要的三要素。此外，还有一些具体的冲泡要求，如泡茶前先温杯、温壶；冲泡后及时做到茶水分离，尤其是在冲泡发酵程度较高的茶时，茶水分离要彻底；用玻璃杯冲泡绿茶时，应在前一泡剩余三分之一左右时续水，以获得较稳定和持续的口感。

In short, tea dose, water temperature and brewing time are crucial to a good cup of tea. There are some other tips for brewing. For example, warm the cup or pot by swirling a small amount of hot water in it before adding the leaves; never keep the tea stand in the pot for too long. When brewing fermented tea, any remaining tea liquid has to be poured out between brews; while brewing green tea with a glass, do not empty the glass completely and leave about one third of the liquid inside to strengthen the next brew.

（4）日常冲泡 / Brewing Tea in Everyday Life

在生活中，茶的冲泡方式会因场合的正式程度不同而有所区别，而茶的冲泡流程、水温控制、时间把控也因人而异、因茶而异。"看茶泡茶"是一名合格的茶艺师应该具备的技能。在此基础上，根据不同场合，结合品茶人的体质、习惯与偏好，"看人泡茶"则是对优秀茶艺师的要求。可以形象地说，泡茶就像做饭，不仅需要好的原料、合适的器皿、高超的烹饪技巧，还需要契合场景，考虑客人的口味偏好。

In daily life, one may find that the way of brewing tea varies with the formality of occasions. Even the same type of tea is sometimes brewed differently. A qualified tea artist should know the difference of various teas and brew them accordingly. A good tea artist is expected to take more things into consideration such as the formality of the occasion and individual preferences or personal tastes. Just like cooking, brewing a good cup of tea depends on good raw materials, appropriate utensils, brewing skills, drinking occasions and preferences of guests.

相对于中国茶叶冲泡的种种技巧与要求，很多国家的茶叶冲泡方式要简单得多。一些国际知名品牌深知冲泡方式的重要性，故为自家产品做了冲泡建议。以川宁为例，它为每种茶标注了最佳冲泡时间和冲泡水温，虽然每个人口味不一，但对于大众口味而言，推荐的时间和温度确实是最适宜的。只需按照推荐步骤操作，一杯可口的茶就触手可及。因此有人认为，欧美人泡茶缺乏审美情趣。其实，对冲泡艺术的理解仁者见仁，茶可口是第一要素。

Compared with the complex requirements of tea brewing skills of China, tea brewing is taken in a much easier way in many other countries. Some famous tea brands have long realized the importance of brewing skills, so they offer simple tips for brewing. If customers like their tea stronger or weaker, they can adjust by themselves. Take Twinings as an example, the optimal time and temperature of brewing the tea are printed on packing boxes, which helps produce a beautiful, smooth cup that caters to most of the customers. What customers should do is to follow the simple recipe. Some people believe that these brewing recipes are more functional than aesthetic. In fact, tastes and preferences vary from individual to individual, and the most important thing is people can make a perfect cup for their own taste.

我们在日常泡茶时，以便捷、舒适为宜，无须墨守成规，纠结于各种规范，只要健康可口即可。在紧张的工作之余，我们可以悠闲地泡上一壶茶，独饮也好，三五好友小聚也罢，细品慢尝，倒也不失为一种便捷且雅致的放松形式。

By the way, when we brew tea in daily life, we do not need 24 implements for the tea preparation, nor have to draw our water from mountain streams. As long as we follow the basic instructions, we may enjoy a cup of healthy and tasty tea. But if you want to have a few moments of relaxation in such a competitive society, brewing a delicate cup of tea and leisurely enjoying it will offer you a chance of escaping from the boredom and stress of everyday life.

得闲泡茶

不同的冲泡方式源于不同的文化，也反映了不同的文化习俗。当我们举杯品饮时，无问西东，只需牢记茶所蕴含的"和而不同"之理念。

Different brewing skills are rooted in different cultures; meanwhile they are also the mirrors reflecting distinctive cultures. In spite of all these differences, we shall keep in mind that east or west, harmony is the best.

3.2 冲泡方式的演变 / The Evolution of Tea Preparation

中国人的饮茶习惯数千年来发生了很大的变化，其演变大致可分为四个主要阶段：茶粥、煎茶^①、点茶和泡茶。不同的品饮方式反映了不同的时代特征。

For thousands of years, the way of preparing tea has undergone many changes. Generally, its evolution can be divided into four stages, namely, the Tea Porridge, the Boiled Tea, the Whipped Tea, and the Steeped Tea, each reflecting the spirit of the age in which they prevailed.

（1）魏晋之前——茶粥 / Till the Wei and Jin Dynasties—Tea Congee

约五千年前，古人就知道野生茶树的功能，他们采摘鲜叶口嚼生吃，药食两用。从先秦至两汉，茶由药用和食用逐步向饮用转变。魏晋时期，茶的食用与饮用并行。初期茶的主要食用形式为将茶的鲜叶与其他食材一起做成粥状饮品。从这种意义上说，彼时的茶可以视为一种特殊的蔬菜，兼具食品和饮品的特性。后来，人们发现茶的鲜叶存贮、运输均不方便，便开始想方设法加工鲜叶，使之干燥成型又不失真味。当时的加工方式是制茶饼^②，饮用方式则是羹饮，即加入米、生姜、盐、橘皮等调料去除苦味，煮成粥状。采茶制饼和混煮羹饮是秦汉至中唐近千年间制茶与饮茶的主要方式。

---

① 此处的煎茶与日本用来代表特定制茶方法或茶叶等级的煎茶是不同的。此处可简单理解为煮茶。

② 三国魏人张揖撰《广雅》云："荆巴间采叶作饼，成以米膏出之。若饮先炙令赤色，捣末置瓷器中，以汤浇覆之，用葱姜芼之。"这是现知有关制作茶饼的最早记录。

About 5000 years ago, Chinese ancestors found the wild tea plant and discovered its therapeutic functions. They picked tea leaves and used the fresh leaves as medicine and food. The leaves were initially chewed rather than brewed or steeped for a drink. From the Qin Dynasty to the Han Dynasty, people began to prepare tea for drinking use, which, in the Wei and Jin dynasties, developed in parallel with tea consumption as medicine and food. At the beginning, thick soup stewed with fresh tea leaves and other ingredients was the mainstream of tea consumption. That is to say, fresh tea leaves were cooked in a similar way as vegetables. However, due to the inconvenience of the storage and transportation of fresh tea leaves, people began to make fresh tea leaves into tea cakes. To cover the bitterness, tea was boiled, together with many additives such as rice, ginger, salt, orange peel and so on. By the middle of the Tang Dynasty, tea had been mainly processed and consumed in this way.

（2）唐代——煎茶 / The Tang Dynasty—The Boiled Tea

唐代，采茶制饼技术愈加精细。鲜叶采摘后，先蒸再碾，置于模具成饼，以炭火焙干后，穿起封存。饮茶时，先将茶饼用火炙烤，而后碾碎，再烧水煮茶，为了改善茶叶的苦涩味，常加入盐、姜等调料[1]，中唐之后除盐之外，其他调料渐少。唐人以釜煮茶，水初沸，加盐；二沸，先舀出一瓢水放置一边，再把茶末投入水中；三沸，倒回冷却茶汤，以助茶汤沫饽[2]。如此，茶便煮好了。

---

[1] 陆羽在《茶经·五之煮》中有"初沸，则水合量，调之以盐味"，即除了盐外，舍弃了其他调料。

[2] 《茶经·五之煮》中有记载"沫饽，汤之华也。华之薄者曰沫，厚者曰饽，轻细者曰花。"

In the Tang dynasty, processing techniques of tea cakes were more sophisticated. When the fresh leaves were picked, they were first steamed, crushed to form a paste, poured into the mold and compressed to form the tea cake. After being roasted, tea cakes were strung for storage. When preparing tea, people first baked the tea cake, and then crushed it into powder. Meanwhile, they boiled some water in a kettle called "Fu". When the water was boiling for the first time, put some salt into Fu; for the second boiling, tea powder was added and a spoonful of soup was taken out; at the third boil, the spoonful of cold soup was poured back into the kettle to settle the tea and revive the "youth of the water". Then the beverage was poured into cups and drunk.

唐代煎茶（龚记永元　供图）

（3）宋代——点茶 / The Song Dynasty—The Whipped Tea

到了宋代，茶品日渐丰富，饮茶也日益考究，人们开始重视茶叶本身的色、香、味，故烹茶时不再使用调味品。宋代风行点茶法，"碾茶为末，注之以汤，以筅击拂"，也就是将茶饼以石磨碾碎，置于碗中，注入少量沸水调成糊状，再分次注入少量沸水，并用竹制的茶筅搅动击打，直至茶水交融，沫饽[①]丰富。因加入热水的动作被称为"点"，点茶由此得名[②]。

In the Song Dynasty, more varieties of tea products had been developed. People attached more importance to tea preparation. At that time, additives were discarded and people gave priority to the color, fragrance and taste of tea itself. A new tea preparing method named "*Dian Cha*", meaning whipped tea came into fashion. The tea cake was first ground to fine powder in a small stone mill; then the fine powdered tea was put into a bowl. After the powder was whipped into paste by a delicate whisk made of split bamboo, more boiled water was added into the bowl for several times. Each time, whip the tea gently and skillfully till the tea was covered with a good head of foam. Called "*mobo*" in Chinese, it is similar to the foam on top of coffee. The action of pouring small amount of hot water into the bowl was called "*Dian*"; hence the Chinese name "*Dian Cha*".

---

①　沫饽看上去与咖啡上的奶盖相似。讲解时，可用咖啡奶盖类比沫饽。

②　中文"点茶"突出逐步加水这个动作——"点"；英文 whipped tea 则强调 whip 这个击打茶汤的动作和技巧。

点茶体验准备（马娜 供图）

（4）明清 —— 泡茶 / The Ming and Qing Dynasties—The Steeped Tea

在饮茶方式上，元代处于一个过渡时期，当然也有人认为元代破坏了前代的茶文化。明代，饼团茶逐渐减少，散茶盛行，烹茶方法以冲泡为主，即用沸水冲泡叶茶。泡茶法是此后600多年中国人的主流饮茶方式。也是在明代，茶传到了欧洲，故用沸水泡茶的方法也传到了西方，并流行至今。

The Yuan Dynasty was in a transitional period in tea drinking and preparation, although some scholars thought it was a disastrous period for tea industry and culture. In the Ming Dynasty, loose leaf tea prevailed for its convenience and original taste. Accordingly, tea preparation was simplified—loose leaf tea was steeped in hot water. Since then, the simple way of steeping tea has been followed by Chinese for more than 600 years. In the Ming Dynasty, tea was introduced to Europe, so the method of brewing tea with boiling water was also introduced to the West and is still popular today.

到了清代，现代所说的六大茶类均已形成，饮茶特点主要体现在茶品和冲泡方法的多样性上。

In the Qing Dynasty, along with the consumption of six types of tea, people have a diversity of different teas and tea preparations varied with different teas.

中国历史悠久、幅员辽阔，各个时期的制茶、饮茶方法并非特定和唯一。同一时期不同区域的冲泡、品饮方式也有区别，很多时候是多种制茶、饮茶方式并存①。以宋代为例，宋人既有制茶饼的，也有做散茶的；既有点茶的，也有煮茶的，甚至还有泡茶的。同样，明代在茶叶以冲泡为主的大环境下，依然留存了不少煮茶和以茶佐饭的习俗记载。

China is a large country with rich culture and a long history. Tea preparations vary with times and regions. Generally, in a certain dynasty or region, there is always a kind of tea preparation prevailing in the society. However, one may also find other tea preparations. Taking the Song Dynasty as an example, although tea was mainly made into cakes in this period, people had already consumed loose leaf tea; and tea was prepared in different ways—whipping, boiling and even steeping. Similarly, in the Ming Dynasty, when tea was mainly brewed with boiling water, there were still many records of the customs of boiling tea and pairing tea with rice.

---

① 陆游在《安国院试茶》诗后注云："日铸则越茶矣，不团不饼，而曰炒青，曰苍鹰爪，则撮泡矣。"由此可知，用沸水冲泡炒青散茶的方法，早在宋代就在民间出现。

（5）当下——便捷多元 / Today—Nothing Is Impossible

今天，茶产业不断创新发展，茶的冲泡和品饮方式更为便捷。20世纪初发明的袋泡茶①，在"二战"之后逐渐流行。袋泡茶发味快又避免了茶渣入口，价格亲民、品质稳定、口味丰富，且一个茶包的茶量刚好可以泡一杯茶，故袋泡茶受到越来越多消费者的喜爱。因传统袋泡茶的便捷性在某种程度上是以品质下降为代价，近年来，为提升品质，袋泡茶可谓"改弦易辙"。金字塔形的设计使茶叶在冲泡时可以完全舒展，而整叶茶加金字塔形茶包的设计让袋泡茶的香气和滋味完全可以媲美整叶茶。

Today, "fast life" is the theme of society. The tea industry is researching and developing new products accordingly. Invented in the beginning of the 20th century, the tea bag got popular after WWII. The tea bag has achieved a dominant position in the world of tea for its convenience, competitive price, reliable quality and a variety of tastes. Besides, a tea bag contains exactly the right amount for a single cup of tea, so wastage is prevented. However, it is generally believed that the taste and aroma of the tea bag can't hold a candle to loose leaf tea. In recent years, tea bags have undergone a complete transformation. Pyramid tea bags consisting of leaf tea guarantee the aroma and taste of tea, since the pyramid shape allows the greatest possible space for tea to "expand" when brewed.

20世纪80年代，罐装茶饮出现；1996年，塑料瓶装茶饮的出

---

① 根据茶叶原料的不同，袋泡茶主要分为绿茶袋泡茶、红茶袋泡茶、乌龙茶袋泡茶、黄茶袋泡茶、白茶袋泡茶、黑茶袋泡茶和花茶袋泡茶。

现①，完全满足了人们随时随地饮茶的需求，自此茶饮料开启了其作为快消品的历程。

In the 1980s, canned tea beverage was invented. In 1996, plastic bottled tea was introduced in Japan, which made it possible for people to drink tea at any time and place. Since then, tea has started its journey as a kind of fast-moving consumer goods.

除了便捷化外，个性化、时尚化亦是当下茶饮的标签。近年来，为更好地保留茶的原真味，智能泡茶机应运而生，人们可以通过选择水温、水量和冲泡时间，制作适合自己口味的茶，只需动动手指就能满足个性化的茶饮需求。

In addition to efficiency of tea consumption, individuality and fashion are also taken into consideration when people drink tea. Intelligent tea brewing

智能泡茶机

machines have been invented in recent years, which may retain the original taste of tea by freshly brewing loose leaf tea. By setting water temperature, tea dose and brewing time, tea can be brewed for individual preference conveniently.

古老茶饮变身时尚先锋。今天，年轻人对以奶茶为代表的新式茶

---

① 引领茶消费方式变化的企业首推为日本的伊藤园，该企业分别于1981年、1985年推出了罐装乌龙茶和罐装绿茶，于20世纪90年代推出塑料瓶装绿茶。

饮的喜爱和追捧成为茶的一大消费热点。新式茶饮以调饮为主，不仅在佐料的搭配上更加丰富，在茶叶的选择上也愈加考究。有创意、口感好、颜值高的新式茶饮成为当下最时尚、最热门的茶叶品饮方式。

As an ancient beverage, tea is consumed in many fashionable ways, among which milk tea (bubble tea/ boba tea) is the most popular. The so called new-style tea beverage is characterized by adding various ingredients. For some big brands, the tea is specially chosen or even custom-made to suit special requirements. New-style tea beverage attracts more and more customers with originality, good taste and delightful appearance.

茶叶品饮方式的变化和制作工艺的变革与创新密切相关。数千年来，茶叶无论在形态还是品饮方式上，都在不断地发生变化。当下丰富的茶叶形态和多样化的品饮方式让人们有了更多的选择，相信大家都会找到自己喜欢的那一杯茶。

To some extent, the innovation of tea processing techniques determines the way people prepare tea. Through thousands of years, tea has been changing in both aspects. Today, with tea processing techniques being greatly diversified, more and more tea products have been developed in response to various demands. Hope you can find your favorite cup of tea!

04 茶之具
Tea Ware

4.1 茶具发展简述 / Evolution of Tea Ware

茶器具的变化总伴随着品饮方式的变化[①]。数千年来，茶具经历了从无到有、由兼而专的过程。茶最早是药用或食用，经历了从生嚼变成羹饮的过程，故彼时并无专门的茶具，而是与食具、酒具共用。当茶逐步成为日常饮料，专用的茶具才逐渐出现。

The history of tea ware always follows the history of tea drinking. Tea ware grew out of nothing and through thousands of years it has experienced a process from simple to exquisite and comprehensive. As we know, when

---

①　陆羽对茶具和茶器做了区分，他在《茶经·二之具》把关于采茶、制茶的工具定义为茶具；在《茶经·四之器》中，把烹茶相关的用具，包括对茶的育化有改善作用的工具，定义为茶器。本章对茶具、茶器未做区分，采用较为常见的"茶具"一词指代泡茶、品茶所需的所有相关用具。

tea was first discovered, fresh tea leaves were initially chewed. Later, it was made into tea congee or soup as a kind of medicine or food, and in that period, wine and food vessels were used for having tea. Utensils specially designed for drinking tea emerged only after tea was consumed as a kind of beverage.

目前已知最早的茶具为浙江湖州出土的一只东汉的青瓷茶罍；最早提到茶具的史料是西汉王褒《僮约》中的"烹茶尽具"。

The earliest physical evidence of tea ware is a Celadon urn-shaped vessel for storing tea from the Eastern Han Dynasty unearthed in Huzhou, Zhejiang Province; and the first textual reference about tea ware is from *Tong yue* by a man named Wang Bao in the Western Han Dynasty.

东晋、南朝时，较精细的饮茶法开始出现，茶具也逐渐从饮食器中分离出来。唐代，茶已成为举国之饮，必然要求与之相伴的茶具有所发展和变革。正是在这一时期，历史上首次体系化的茶具出现了，这就是陆羽在《茶经》中列出的 20 多种彼时烹茶所需的器具。陆羽描述的茶具形制明确，兼具实用性和审美性，同时蕴含茶文化精神。

During the Eastern Jin and Southern Dynasties, tea was consumed more as a beverage than as a soup, which required utensils or accessories designed for tea drinking. When tea became the national drink in the Tang Dynasty, the first set of tea ware emerged. It was Lu Yu who listed more than 20 implements that were essential for the correct preparation of a cup of tea. The implements were clearly and vividly described and illustrated, combining both functional and aesthetic values. More importantly, the spirit of tea culture was embodied in these implements.

在唐代"南青北白"（即南方以越窑为代表的青瓷和北方以邢窑为代表的白瓷）的瓷器格局下，实用与审美并重的瓷质茶具不断涌现。陆羽认为"邢不如越"，主要因为"青则益茶"，即青色会为茶汤增添一些迷人的色泽，而白色会让茶汤看起来偏红，让人没那么有胃口[①]。除此之外，颇受皇家所爱的金属茶具同样璀璨夺目，尽显大唐风范。

In the Tang Dynasty, celadon in the south and white porcelain in the north were favored because they matched well with the boiled tea. Lu Yu considered the greenish blue of celadon as the ideal color for the tea-cup, as it lent additional greenness to the beverage, whereas the white made it look pinkish and distasteful. Apart from delicate porcelain tea ware, the dazzling gold and silver tea ware was also favored, mainly by the royalty.

唐代宫廷茶具（静观　供图）

宋代点茶法盛行，故对茶具的形制、材质、色泽要求与煎茶时期不同。总体而言，宋代茶具在数量上要比唐代少一些[②]，但在制作工艺、

---

①　如陆羽所言"若邢瓷类银，越瓷类玉，邢不如越，一也。若邢瓷类雪，则越瓷类冰，邢不如越，二也。邢瓷白而茶色丹，越瓷青而茶色绿，邢不如越，三也"。

②　白堃元教授认为宋代茶具在以下四个方面有变化：一是改碗为盏；二是改鍑为瓶；三是改竹夹为茶钤；四是改栟榈为茶筅。

审美情趣上，比唐代更加精细多姿。人们开始偏好黑色或深褐色的厚碗，因为这些颜色能凸显点茶汤花的纯白，而厚碗则有利于蓄热，便于茶沫饽的呈现。当时产于建安的黑釉盏（建盏）颜色、材质都能体现点茶的精髓之所在，故大受追捧，格外珍贵[①]。宋代点茶用具的集大成者非南宋审安老人的《茶具图赞》莫属。作者用白描画法画出了宋代点茶所用的十二件茶具，称为“十二先生”，并按茶具特点冠以姓氏与宋代职官名号，生动反映了十二种茶具的材质和功用[②]。这是我国现存最早的茶具图谱。

In the Song Dynasty, people took to the powered tea so tea implements were different from those used for boiled tea in shape, texture and color. Generally, there were fewer utensils used for whipping tea; however, they were more exquisite and functional than those of the previous dynasties. People preferred heavy bowls of black and dark brown. It was believed that black was the ideal color for the tea bowl since it could highlight the pure white of the tea foam and the heavy bowl could make the tea foam last longer which was a key criterion of a good bowl of whipped tea. The most popular tea bowls at that time were black glazed tea bowls made in Jian'an (now Jian'ou, Fujian province). The greatest collection of tea ware of that time can only be found in “Illustrated Catalogue of Tea Ware” written in the Southern Song Dynasty. All together 12 tea implements were vividly painted and each was given a surname and an official title. The most interesting thing is that each surname and official title is carefully chosen to show the

---

① 如蔡襄所言“茶色白，宜黑盏，建安所造者绀黑，纹如兔毫，其胚微厚，熁之久热难冷，最为要用”。

② 十二件茶具分别是：韦鸿胪（茶笼）、木待制（木椎）、金法曹（茶碾）、石转运（茶磨）、胡员外（茶杓）、罗枢密（茶罗）、宗从事（茶帚）、漆雕秘阁（茶托）、陶宝文（茶盏）、汤提点（汤瓶）、竺副帅（茶筅）和司职方（茶巾）。

宋代点茶"十二先生"

function and material of each tea ware. It is the earliest catalogue of tea ware in Chinese history.

　　元代在茶具发展上处于过渡期。明清时期，饮茶的主流方式为散茶冲泡法，饮茶方式的改变带动了茶具的变革，出现了小茶壶等适合冲泡的茶具。景德镇瓷茶具和宜兴紫砂茶具风行大江南北。景德镇瓷器创新不断，至精至美，一枝独秀[①]，同时，被誉为实用与艺术完美结合的紫砂茶壶登上历史舞台，与瓷器争名。用紫砂壶泡茶，香浓味正，备受追捧。总体而言，明代茶具简约雅致，清代茶具精细丰富。

　　The Yuan Dynasty was in a transitional period in the development of tea ware. During the Ming and Qing dynasties, people took to the steeped tea, and small teapots emerged, which led to the popularity of porcelain tea ware made in Jingdezhen and purple clay tea ware in Yixing. Purple clay teapots,

---

　　① 中国瓷器生产发展从唐、宋时代众多窑口"百花争艳"的态势，经由元代过渡以后，到明代基本演变为景德镇窑"一枝独秀"的局面，景德镇被誉为"天下窑器所聚"之地。明、清两代，精美的瓷器多出于景德镇。

the perfect combination of practicality and artistry, were particularly sought after since they could enhance the taste and fragrance of tea. Comparatively, tea ware of the Ming Dynasty was simple and elegant, while that of the Qing Dynasty was refined and diverse.

今天，饮茶的主流方式延续明清，故茶具的种类没有太大的变化，但茶具在材质、功能和造型上都有更多元的表现与发展。

Today, people mainly follow the tea preparation of steeping tea in hot water, so tea ware is generally similar to that in the Ming and Qing dynasties. However, if you want to try something new, something special and something unique, there are plenty of innovative and exotic tea utensils or accessories available.

茶具的产生和发展既与社会经济文化有关，也与时代习俗、审美情趣以及茶叶制备、饮茶方法的变化有关。每个历史时期的茶具在一定程度上都反映那个时期的技艺水平和精神价值。

The emergence and development of tea implements are related to social, economic, and cultural factors. They are the showcase of changes in customs, aesthetic values as well as methods of tea preparation and drinking of a certain period of time. The history of tea ware is the epitome of development of the craftsmanship and spiritual value of tea.

4.2 茶具分类 / Classification of Tea Ware

茶具通常按功能和材质分类。从功能上看，茶具可分为饮具、煮具、贮具、碾碎具、洁具、水具和辅具等；从材质上看，茶具可分为陶瓷茶具、金属茶具、竹木茶具、玻璃茶具、漆器茶具和塑料茶具等 ① 。

Tea ware is usually classified according to the function or the material. In terms of functions, there are utensils for drinking tea, for boiling water, for storing tea, for grinding tea and so on. In terms of materials, there are utensils made in ceramic, metal, bamboo and wood, glass, lacquer, plastic and so on.

陶瓷茶具是最常用的茶具，按材质可进一步分为瓷质和陶质两种。瓷质茶具主要有青瓷、白瓷、黑瓷和彩瓷。青瓷以浙江生产的为最好，其中釉色温润、工艺精湛的龙泉青瓷有"雪拉同"的美称；白瓷则以福建德化生产的为佳，洁白无瑕的颜色最能反映茶汤；黑瓷茶具鼎盛于宋代，以产于福建建安（今福建南平建阳区）的建盏为最佳；在彩瓷茶具中，青花瓷在元代中后期渐成气候，数百年来，人们对它的偏好经久不衰。

Ceramic tea ware is the most commonly used utensils, which can be further divided into porcelain tea ware and pottery tea ware. Porcelain tea ware is in rich colors and patterns. Longquan, Zhejiang Province is known for a type of greenish or grayish blue ceramics called celadon, which has long been favored by people for its enchanting glazing color and exquisite craftsmanship. White porcelains produced in Dehua, Fujian Province are

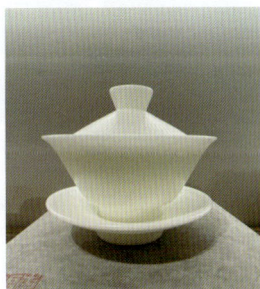

德化白瓷

---

① 历史上还有用玉石、水晶、玛瑙等材料制作的茶具，但这些茶具数量较少，且以美学价值为主，在茶具史上居次要地位，故在此不做讨论。

of good quality, in which tea liquor looks clearer and more tasteful. Black glazed porcelain tea ware flourished in the Song Dynasty, with those produced in Jian'an (now Jian Yang in Nanping, Fujian Province) being the best. Among various colored porcelain tea ware, the blue and white porcelain tea ware became popular in the mid to late Yuan Dynasty, and since then it has remained one of the most popular tea ware for hundreds of years.

陶质茶具中以宜兴紫砂茶具最为出名。宜兴紫砂茶具（以紫砂壶为代表）的原料为紫砂泥，产自江苏宜兴，故得名。紫砂壶始于宋代，盛于明清，流传至今。紫砂壶透气性好，能够激发茶性，使茶汤更加醇厚。同时，紫砂壶艺术价值高且把玩性强。紫砂壶的透气性和吸附性都很强，正因如此，紫砂壶的清洗养护要求也比较高。紫砂壶的缺点[①]是通用性不高，不适合冲泡细嫩的绿茶，故最好按不同茶类分别选择紫砂壶。

紫砂壶

Among tea ware made of pottery, Yixing purple clay tea ware is the most precious, which is widely admired for its functional and artistic merits. The tea ware (mostly teapot), made with a kind of special clay called purple sand (Zisha), is from the area of Yixing, in Jiangsu province. The clay was first mined during the Song Dynasty. In the Ming and Qing dynasties, it became the preferred teapot of refined scholars. The purple clay teapot can absorb

① 因为紫砂壶较高的艺术价值和原料的稀缺性，使市面上紫砂壶精品的价格较高，导致劣等仿品充斥市场，不仅危害健康，还造成紫砂市场的乱象。

the fragrance and taste of tea during each brew so some tea fanatics will use one teapot for one type of tea only. For this reason, purple clay tea pots need careful maintenance. Besides, some delicate teas such as tender green tea cannot be well brewed with purple clay teapots.

金属茶具包括由金、银、铜、铁、锡等金属材料制成的器具。隋唐时期是金银器具制作的高峰，出土于陕西省宝鸡市扶风县法门寺的一套皇家鎏金茶具是金属茶具中的稀世珍宝。明代随着饮茶方式的改变，金属茶具已渐少。但用银、铁、锡等材质制作的煮水壶、茶叶罐等则既具有艺术价值，又有很高的实用性，颇受茶人喜爱。

Metal tea ware refers to utensils made of gold, silver, copper, iron, etc. As early as the Sui and Tang dynasties, gold and silver tea ware was sophisticated in technology. The highlight or the typical example of gold and silver tea ware is a set of royal gilded tea ware, unearthed from Famen Temple, located in Fufeng county of Shaanxi province. With the change of tea preparation in the Ming Dynasty, metal tea ware was gradually replaced by ceramic one. However, silver or iron water boilers, and tin tea caddies with both artistic and practical values are still greatly favored by tea lovers.

铜质水盂

日式铁壶

竹木茶具在唐代以前使用较多。竹木茶具的优点是原料易得，制作方便；缺点是易发霉腐烂。当下，竹木茶具中以竹编茶具最受人们喜爱。竹编茶具通常由内胎和外套组成，内胎多为陶瓷材质，外套为精致竹编。

Bamboo and wood tea ware was widely used before the Tang Dynasty, with the advantages of availability of raw materials and convenient production. The disadvantage of these utensils is that they mold and decay easily. Today, bamboo woven utensils are quite popular which are different from the traditional ones in design. They usually consist of two layers, with the inner layer mostly made of ceramics and the outer layer made of exquisite bamboo weaving.

竹编茶具

防烫玻璃茶杯

玻璃质地透明、形态各异、用途广泛、价格亲民。在玻璃杯中泡茶，可以观赏茶叶在水中的舒展沉浮和茶汤颜色，在夏天的茶席上，还可以增加一丝清凉感。玻璃器具的缺点是容易破碎，也比较烫手，且无法协助表达一些强劲的好茶的优点。

Glass tea ware is transparent in texture, diverse in shape, and affordable in price. Besides, it is easy to use and easy to clean. So it is widely used in daily life. Brewing tea in a glass utensil allows people to observe the tea leaves stretching and floating in the water. However, glass tea ware breaks easily and it is sometimes too hot to be handled. Besides, glass tea ware cannot reveal the strength of robust teas.

漆器茶具主要产自福建福州。漆器茶具轻巧美观、色彩艳丽夺目，总体而言其艺术性高于实用性。塑料茶具色彩丰富、携带方便、价格低廉，但因其可能存在的健康隐患，建议谨慎选用。

Lacquer tea ware is mainly produced in Fuzhou, Fujian province. They are light and beautiful, with bright colors. People choose them mainly for the artistic purpose rather than functional. Plastic tea ware is colorful, easy to carry, and inexpensive, but it may pose potential risk to our health. Therefore, we'd better not use them too often.

各式漆器茶具

4.3 茶具简介 / Brief Introduction to Tea Ware

（1）主要茶具 / Basic Tea Ware

按茶具的用途和存在的意义，可将茶具分为主要茶具和辅助茶具。主要茶具包括煮水壶、茶壶、盖碗、公道杯、品茗杯等。

Tea ware can also be divided into primary or basic tea ware and auxiliary tea ware according to the importance and functions in brewing. The basic tea ware includes the water boiler, the teapot, Gaiwan, the fair cup, the tea cup, etc.

煮水壶： 不锈钢壶、陶壶、玻璃壶、铁壶、银壶等[①]。

茶壶： 用于泡茶，材质多样，造型丰富。

盖碗： 用于泡茶或个人品茶[②]。

银质煮水壶

汝窑茶壶

盖碗

---

① 煮水方式一般包括用电、酒精和各类炭。

② 盖碗具有灵活性和多样性的特点。冲泡饮用皆可，既能赏茶、品茶又能闻香。盖碗通常是高温瓷，不易吸附味道，故一个盖碗可冲泡多种茶类。盖碗的容量一般在100~180毫升。一人饮茶时需减少投茶量，盖碗还可作大茶杯使用。盖碗是适合大多数人的泡茶器具。但盖碗也有缺点，如盖碗表达不出一款特别优秀老茶的极致口感；此外，对初学者而言，盖碗泡茶也容易烫手。

公道杯：均匀茶汤，便于分茶①。

品茗杯：用于喝茶，材质多样，造型丰富。

闻香杯：用于闻茶香②。

茶盘／茶船：用于盛放茶具，便于泡茶品茶③。

锡制公道杯　　　　瓷制公道杯　　　　各式品茗杯　　　　锡制茶船
（马娜　供图）

茶叶罐：用于储存茶叶。

茶巾：用于擦除水渍、清理茶桌（通常材质为吸水性较好的棉麻）。

茶滤：用于过滤茶渣，使茶汤更清澈，通常置于公道杯上。

水盂：用于盛放茶渣和泡茶过程中产生的废水。

Water boiler (or tea kettle): used for boiling water, usually made of stainless steel, ceramics, glass, iron or even silver, etc.

Teapot: used for brewing tea, coming in different sizes and materials, plain or decorated, modern or vintage.

Gaiwan: a three-piece tea set, used for brewing tea or drinking tea.

---

① 公道杯可以及时进行茶水分离，有利于控制茶汤的滋味。在生活中泡茶，英式下午茶茶具中的奶盅也可以用作公道杯。

② 闻香杯多为瓷质，与同材质品茗杯一起使用，比品茗杯细长，利于聚香，多用于冲泡台式乌龙茶。

③ 茶盘亦可指奉茶之用的平板状托盘，有些茶人用茶船指代公道杯。此处茶盘／茶船，指泡茶时放置茶具并可承接废弃茶水的器具。

Fair cup (or fair mug): used for balancing the thickness of tea liquor and evenly distributing tea liquor.

Tea cup: used for drinking tea, coming in different sizes and materials.

Aroma-smelling cup: used for better smelling and appreciating the aroma of tea (mainly for oolong tea).

Tea tray/ Tea ship: used for holding tea utensils on which the whole brewing process can be performed.

Tea canister (tea caddy): used for storing tea.

Tea towel: used for wiping off water stains and cleaning the tea table.

Tea filter (tea strainer): used for filtering tea leaves when pour tea into a cup, to make the tea liquor clear and pure, usually placed on the fair cup.

Water basin: used for holding waste water or tea dregs during the brewing process.

茶叶罐与茶巾（云晴　供图）　　　　银质茶滤　　　　　　　水盂

（2）辅助茶具 / Auxiliary Tea Ware

辅助茶具首先包括"茶道六君子"。顾名思义，"六君子"一般为六件，通常包括茶筒、茶漏、茶夹、茶匙、茶针和茶拨，该组套器具材质多为竹木或金属。

The most important auxiliary tea ware is the "Six Treasures of Tea Ceremony", which is usually made of bamboo, wood or metal. As the name

suggests, the "Six Treasures" consist of six pieces, including a tool holder, a tea funnel, a tea tong, a tea spoon, a teapot spout needle, and a tea stick.

茶筒：用于收纳组合中的其他茶器具。

茶漏：壶口较小的时候，可以将茶漏置于壶口，便于茶叶入壶。

茶夹：用于夹杯子、辅料等，防烫又卫生。

茶匙：用于优雅又卫生地从茶叶罐或包装袋里取茶叶。

茶针：用于疏通壶嘴或拨动茶叶。

茶拨：用于将茶叶由茶荷拨入壶中[①]。

Tool holder: home to the other 5 tools or used for keeping the other 5 tools.

Tea funnel: preventing tea leaves from falling out of the teapot, or through which tea leaves can be easily put into the teapot by enlarging the area of the mouth of the teapot.

Tea tongs: used for picking up teacups or tea dregs.

Tea spoon: used for taking tea leaves from the tea caddy in an elegant and hygienic way[②].

茶道六君子

Tea Spout needle: used for dredging the teapot spout.

Tea stick: used for putting tea leaves into the tea pot.

[①] 因茶拨的形制和功能与茶针比较接近，目前也有一些茶道组合将茶拨替换为撬茶用的茶锥。另外，因泡茶所用的辅助工具很多，故在工具名称上，不同的书籍有不同的命名和解释。

[②] 英式下午茶中的茶匙指饮茶时搅茶用的小勺子（Tea spoon refers to the small spoon for stirring tea.）。

此外，还有其他一些常见的辅助茶具，如茶荷、壶承、杯垫、茶刀、盖置、茶宠、养壶笔等。

Other auxiliary tea ware or tea accessories such as the tea holder, the teapot holder, the coaster, the tea knife, the lid holder, the tea pet and the teapot brush also help make a good cup of tea.

茶荷：用于盛放待泡干茶的器皿，可展示、观赏茶叶，也便于控制投茶量（通常与茶拨搭配使用）。

壶承：用于放置茶壶，既能隔热，防止茶壶烫伤桌面，又能承接水渍，保持茶席整洁美观。

杯垫：用于放置茶杯。

茶荷

壶承（丹 供图）

杯垫

茶刀/茶锥：用于撬茶、拆茶饼[1]。

盖置：用于放置壶盖。

茶宠：用茶水滋养的宠物，用以增添品茗情趣[2]。

养壶笔：用于清理茶盘，养壶、养茶宠。

---

[1] 撬茶、拆茶是个技术活，撬得不好可能会伤到手。完整的条索泡出来的茶汤味道更好，故撬茶水平的好坏，也会影响茶的口感。Breaking the tea cake might be a dangerous thing, which needs a skillful hand. Remember whole leaves make better tea.

[2] 茶宠多为紫砂或澄泥烧制的陶质工艺品，也有一些为瓷质或石质。其常见形象包括金蟾、貔貅、小动物、瓜果蔬菜等。通过日常的滋养和把玩，茶宠会变得温润光滑，让人赏心悦目。

Tea holder: used for measuring out the tea leaves and allowing the tea leaves to be clearly viewed before scraping them into the brewing vessel.

Teapot holder: used for keeping the brewing process clean and hygienic and protecting the top of a table; put underneath a teapot.

Coaster/cup mat: used for placing tea cups to protect the top of a table.

Tea knife/Compressed tea needle: used for breaking up the tea cake into lumps of smaller piece[①].

Lid holder: used for holding the lid of the brewing utensil.

Tea pet: small classical Chinese figurines over which tea liquor is poured, used for adding fun to the whole drinking experience.

Teapot brush: used for keeping teapots and tea trays clean, "raising" tea pets.

茶锥　　　　　　　　　　盖置　　　　　　　　　螃蟹茶宠

"工欲善其事，必先利其器"，选择合适的茶具可以更好地展示和体会茶的意韵。换句话说，茶的美也存在或依赖于服务它的载体，小到一个品茗杯，选得合适，也可以帮助我们营造美好的品茶体验。选茶具一般考虑两个因素，一是功能，即茶具的实用性；二是美观，即

---

① 为与茶道六君子中的茶针区别，此处将茶刀/茶锥翻译为常用的 tea knife 或 compressed tea needle。

茶具的艺术性。我们可以首先根据所冲泡的茶类选择合适材质和造型的茶具，在此基础上，本着"应季""应景""应情"的原则，再选择赏心悦目的茶具组合。当然，日常泡茶可繁可简，需要的茶具也是可多可少，一个盖碗或是玻璃杯，只要用得贴切，也可以让我们享受到美妙的品茶片刻。

As an old Chinese saying goes "Sharp tools make good work", the proper tea ware helps enhance the essence of tea. People generally take two factors into consideration when choosing tea ware. Firstly, the tea set should match the tea; that is to say, its material, shape, size and function can help brew a perfect cup. Secondly, all tea utensils should match up with each other to create an enchanting overall effect. Besides, tea utensils can be chosen according to different seasons, occasions and purposes. A good cup of tea does not always require a set of exquisite tea utensils. As long as we make an effort to brew a good cup with proper tea ware, we may enjoy a moment of wonderful tea time.

05 茶之饮
Tea Drinking

5.1 茶品饮 / Tea Appreciation

　　茶遇水绽放，以姿态万千、香气四溢、滋味无穷的方式重生。品茶是对茶的色、香、味、形等各方面的观察、评价和享受。在这几个因素中最为人们关注的是茶的滋味。据研究，茶的苦味主要来自茶叶中的咖啡碱、花青素及茶皂素等；涩味则主要源于多酚类物质；鲜味的主体是氨基酸；甜味源于糖类和部分氨基酸物质；酸味则与有机酸和茶黄素等有关。

　　Tea leaves bloom in water in a myriad of shapes, fragrance and flavors. Tea tasting is the observation, evaluation, and enjoyment of various aspects of tea including its color, aroma, taste, and shape. Among these factors, the taste of tea is the main focus of attention. Researchers have found that bitter taste of tea mainly comes from caffeine, anthocyanins, and tea saponins;

the astringency is mainly from polyphenolic substances; the umami flavor is from amino acids; sweetness is from saccharides and some amino acid substances; sour taste is related to organic acids and theaflavins.

在数千年的历史长河中，茶早已融入中国人的日常生活。大多数中国人喝茶，不添加任何辅料，关注的是茶原本的色、香、味。

For thousands of years, tea has been an indispensable part of the daily life of Chinese people. The essence of the enjoyment of tea lies in the appreciation of its color, fragrance and flavor. Therefore, most of us prefer plain tea—without adding anything.

日常雅事（马娜　供图）

对中国人而言，茶是"等同米盐，不可一日以无^①"的每日必需品，也是"不许戾家"的"四般闲事"之一，与烧香、挂画、插花共列四大雅道^②。换句话说，茶既是解渴的日常饮品，也是清雅的文化载体。

For us Chinese, tea can be as ordinary as rice and salt. Tea drinking, meanwhile, can also be as elegant as the graceful leisure

① 语出王安石《议茶法》。
② 语出宋代耐得翁《都城纪胜》，原文为"烧香点茶，挂画插花，四般闲事，不许戾家"。此处"戾家"，指外行人；"不许戾家"指不要去问外行人，即要托付给专业的人去做。

activities like incense burning, painting appreciation, and flowers arranging. In other words, tea has so many possibilities that it can almost cater for the needs of everyone.

(1) 品茶五美 / Five Aspects Essential for the Appreciation of Tea

君子小人、富贵贫贱皆可饮茶。寻常百姓日常饮茶，重在便捷，以解渴为主要目的。但对文人雅士而言，品茶乃生活之乐，因此他们对品茶的细节极为考究。宋代文豪欧阳修认为品茶必须茶新、水甘、器洁，再加上天朗、客嘉①，此"五美"俱全，方可达到"真物有真赏"的境界。

Although tea is for everyone, the function and the way of drinking tea vary from person to person. For ordinary people, tea is a necessity of life so it is often conveniently prepared mainly to slake one's thirst. While for scholars and gentle men, tea drinking has come to be considered more of an art than a daily routine. Therefore, it is not surprising that they are very demanding when enjoying tea. According to the famous poet Ouyang Xiu in the Song Dynasty, the proper enjoyment of tea can only be developed when there are freshly-made tea, sweet water, clean tea ware, good environment and graceful guests.

欧阳修认为的"茶新"是好茶的第一要素，但考虑到今天我们有丰富的茶类可供选择，我们可将其理解为选择合适的茶品。在生活中，

---

① 欧阳修的这段品茶心得源自《尝新茶呈圣俞》，原句为："建安三千里，京师三月尝新茶……泉甘器洁天色好，坐中拣择客亦嘉。"

茶的选择既有对茶的品质的客观要求，也有个人喜好、时令节气等多方面因素，可以说是因人而异、因时而异、因事而异。

According to Ouyang Xiu, "freshly-made tea" is the criterion for good tea. However, there are a wide range of teas available today, and freshness is not the only requisite for good tea. Therefore, in daily life we may try to figure out what is the "right tea". The "right tea" should be of good quality; besides, we should also consider factors like personal preferences, seasonal factors and so on. Choosing the so-called "right tea" is not easy, for it varies from person to person, from time to time, from occasion to occasion.

在"五美"之中，茶、水、器的品质和选择是决定茶滋味的主要因素，也是冲泡一杯好茶的基本要素[①]。除此之外，良好的品饮体验还需要有好的天气，即茶应在宜人的天气下细品慢尝。如明代许次纾就曾提出"饮时……风日晴和、轻阴微雨"。

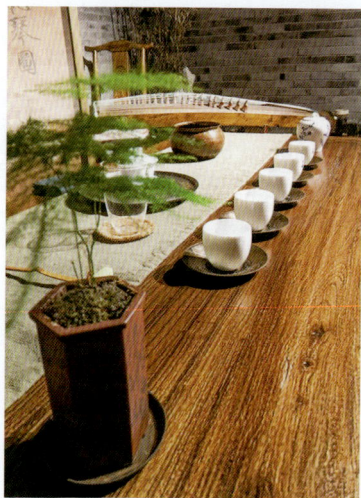

茶席一隅（马娜 供图）

Among the five elements, high quality tea, freshly boiled water and suitable tea ware are essential for a perfect cup of tea. Besides, fine weather is also an indispensable element for a pleasant drinking experience, as Xu Cishu, an expert on tea in the Ming Dynasty, once wrote: "Proper moments for drinking tea…

---

① 择水、备器此前已有论述，在此不再赘述。

When the day is clear and the breeze is mild; on a day of light showers."[①]

"五美"的最后一个要素是"客嘉",即茶还要与具有相当审美修养和情趣的人结伴共赏。林语堂以为,若要品得杯中滋味,客不可多,且须是雅致之人。在美好的氛围中,与志同道合的友人一同品茗聊天无疑是人生一大乐事。

In addition, a perfect tea drinking experience also requires guests of the same type of temperament. According to Lin Yutang, for appreciation of rare cups, one must have quite friends and not too many of them at one time. In such a congenial atmosphere, chatting over a cup of tea with friends is a true enjoyment of life.

(2)影响品饮的其他要素 / Other Aspects Essential for the Enjoyment of Tea

"五美"之外,古人也非常重视品饮环境,即饮茶要有清风明月、茂林修竹等美景营造氛围。许次纾就提出"饮时……小桥画舫、茂林修竹",他认为不宜饮茶的环境包括"阴室、厨房、市喧、小儿啼、野性人、童奴相哄、酷热斋舍"。今天,人们对品饮环境要求虽不似古时那般讲究,但室内饮茶,仍要求窗明几净、清新雅致;户外品饮,则强调在自然美景中感受"人在草木间的乐趣"。

The environment in which tea is tasted is another important aspect. Ancient people believed that tea should be enjoyed in fair weather, with a cool breeze and a bright moon overhead, and lush trees and bamboo growing nearby. According to Xu Cishu, tea should be drunk "in a painted

---

① 译文采用林语堂在 *The Importance of Living* 中的表述。

boat near a small wooden bridge or in a forest with tall bamboos"; he also pointed out that things and places to be kept away from when enjoying tea included damp rooms, kitchens, noisy streets, crying infants, hotheaded persons, quarreling servants, and hot rooms. Today, when people enjoy tea indoors, they are expected to keep the room clean and clear; when outdoors, they usually choose scenic places in which they may feel the harmony of humanity and nature.

户外品饮

　　茶需静品，故很多人认为，品茶以人少为宜，有道是："饮茶以客少为贵，客众则喧，喧则雅趣乏矣。独啜曰幽，二客曰胜，三四曰趣，五六曰泛，七八曰施①。"

　　Tea is invented for quiet company. The company, therefore, must be small. A tea expert in the Ming Dynasty once wrote "it is important in drinking tea that the guests be few. Many guests would make it noisy, and noisiness takes

----

　　① 语出明代张源《茶录》。译文采用林语堂在 The Importance of Living（《生活的艺术》）中的表述。

away its cultured charm. To drink alone is called secluded; to drink between two is called comfortable; to drink with three or four is called charming, to drink with five or six is called common; and to drink with seven or eight is called philanthropic."

都市茶馆（黄佳蕴 供图）

　　在品茶的量上，也是仁者见仁。皎然三饮得道，卢全七碗成仙，两位大师每一次品饮，都有层层递进的身心体验。然而曹雪芹借由《红楼梦》中的妙玉之口，对品茶之量，有不同的见解："一杯为品，二杯即是解渴的蠢物，三杯便是饮牛饮骡了"。

　　There are different opinions concerning the amount of tea one should drink. The famous monk in the Tang Dynasty, Jiao Ran, became immortals when he drank three bowls of tea, while Lu Tong in the Tang Dynasty achieved immortality when he drank seven. For them, each cup brought different life experience. However, Miaoyu, a character in the novel "*A Dream of Red Mansions*", had a unique insight on the amount of tea one should drink, "the first cup of tea is for those who enjoy tea; the second cup is for the philistine to quench their thirst; and the third cup is for those who are too vulgar to be human".

　　茶韵悠长，品茶方式亦有讲究，若要品出茶的精髓，我们应遵循"一看、二闻、三品、四感受"的品茶方法。品的时候不要一口喝完，相反要小口啜饮。此外，虽说提倡吃饭时不要发出声音，但在品茶时，我们会让茶汤在口腔中充分回旋，并发出声音。据说如此可以让氧气与茶汤更好地结合，有助于品味茶汤。

There is a subtle charm in the taste of tea which makes it irresistible. When we are drinking tea, we look for four key features: appearance, aroma, flavor and mouth feel. Tea should never be downed in one gulp; instead, we should drink it slowly at small sips. When drinking tea, forget all table manners and just slurp the tea up into your mouth from the surface of the cup. The louder the slurp, the better you will appreciate the essence of the tea. It is believed that by doing so, oxygen may be mixed with the liquor, which may help people appreciate the flavor.

另外，品茶也需掌握"品"的时机。这里的时机，一是指不同茶的最佳赏味期。传统中医认为"天有五行，人有五脏"，喝茶也该顺应天地自然的变化，应季节而养，循五脏而补，故春季是品饮绿茶的好时节，而秋季乌龙茶则更为适宜。二是指茶需现泡现饮，通常好茶是

夏日茶席

不能沤在壶里的，否则很容易失了真味 ①。

In terms of good timing of tea drinking, it is suggested that different teas should be enjoyed at different times to achieve the best effect. Traditional Chinese medicine believes that drinking tea should follow the course of nature, so green tea should be enjoyed in spring while oolong tea tastes best in autumn. Besides, tea should be freshly brewed and enjoyed timely. If good tea is expected, it should not be allowed to stand in the pot for too long, when the infusion has gone too far.

5.2 泡饮同趣 / Preparation Being Half the Fun of Drinking

值得一提的是，泡茶与品茶二者密不可分。对很多人来说，品茶的乐趣一半在泡，一半在品。鉴于泡茶技艺对茶汤品质的影响，为烹出一口好茶，很多茶人都亲自动手煮水烹茶。在这一点上，林语堂有一段精辟的论述：

真正的茶人享受亲自烹茶的过程。不同于有严谨规程的日本茶道，中国的烹茶品茶一直是富有乐趣、郑重且独特之事。品茶之妙一半在烹茶，恰如嗑瓜子的一半乐趣就在"嗑"这个过程。

It is noteworthy that the preparation and enjoyment of tea are inextricably linked. The process of brewing is part of the enjoyment of tea. Out of consideration for the utmost rightness in tea preparation, many

---

① 有些茶人在冲泡老白茶时会选择闷泡，即将茶叶置于闷泡杯，泡上少至几分钟，多至数小时。大部分茶不宜闷泡，闷泡的茶不仅苦涩味重，还会析出一些有害物质。

connoisseurs of tea have always insisted on personal attention in brewing tea. Lin Yutang captured the essences of tea enjoyment, as he wrote:

> *True connoisseurs regard the personal preparation of tea as a special pleasure. Without developing into a rigid system as in Japan, the preparation and drinking of tea in China is always a performance of loving pleasure, importance and distinction. In fact, the preparation is half the fun of the drinking, as cracking melon-seeds between one's teeth is half the pleasure of eating them.* [①]

品茶之意境（黄佳蕴 供图）

对茶、水、器的选择，对天气、环境、客人、时机的追求，对品茶的量、人数、方式的考究，凡此种种，无不蕴含品茶所追求的精神意境——"和"。

As discussed above, the enjoyment of tea lies in a variety of factors, including the choice of tea, water and utensils, the pursuit of good weather, beautiful environment, graceful guests and good timing, and the consideration of the amount of tea one can drink, number of guests and the proper way of tea tasting. All these show the ultimate goal of Chinese tea culture: the harmony of humanity and nature.

面对这么多品茶的讲究，我们无须对标照搬。在日常生活中，我

---

① 语出林语堂 *The Importance of Living*，中文为笔者所译。

大自在（静观 供图）

们只需秉持泡好一壶茶的心愿，根据实际情况，合理安排即可。无论是饭后家人围坐一桌，共品一壶烟火气十足的聊天茶，抑或是山间湖边，一众志同道合的友人，借清雅之茶，指点江山，只要能品出杯中期待的滋味，心生欢喜，便是当下最好的一杯茶。

In daily life, we do not need to observe all these requirements or restrictions. All we should do is to capture the "flavor" of the moment — the spirit of the season, of the occasion, of the time and the place and cherish each cup of tea. Whether it is a leisure one after a family gathering or a graceful one with friends at woods or by the lake, it is the best cup of tea as long as we enjoy that moment.

小贴士：

品茶时的仪态也需注意。比如很多人用惯了马克杯、玻璃杯，不会用传统的盖碗。梁实秋在《喝茶》一文中曾说"……再不就是在电视剧中也常看见有盖碗茶，可是演员一手执盖一手执碗缩着脖子啜茶那副狼狈相，令人发噱，因为他不知道喝盖碗茶应该是怎样的喝法"。在跟外国友人介绍用盖碗饮茶时，我们可以如下

讲解并示范:

喝茶时应坐姿端正,将茶碗送到嘴边,无须拿掉茶盖,但要将茶盖倾斜以便于饮茶,也可用茶盖将茶叶刮到一旁后再品饮。不能用嘴吹飘浮的茶叶,更不可将吃到嘴里的茶叶再吐回盖碗。此外,在冲泡和品饮时,尽量不发出器具碰撞之声。

The authentic way of drinking Gaiwan tea is to hold the tea saucer in your left hand, pinch the lid with your right hand while gently pushing the tea leaves in the bowl aside with the lid, and then drink the tea by tilting the tea bowl toward your mouth.

第三部分
茶养生气
Better Tea,
Better Health

视觉中国 供图

06 茶之藏
The Storage of Tea

6.1 茶叶贮存的基本原则 / Basic Principles of Tea Storage

（1）茶叶贮存的重要性 / The Importance of Proper Tea Storage

茶叶贮存包括以下几个阶段，从毛茶到精制茶的加工阶段、从工厂到店铺的运输阶段、店铺贮存阶段和家庭贮存阶段。良好的茶叶贮存不仅可以保护消费者的权益，还可以充分发挥茶叶的经济价值，提高茶业的经济效益。作为消费者，我们主要探讨如何在家庭贮存阶段，让茶叶尽可能地保鲜保质。

Tea storage consists of a number of stages, including the processing stage from primary tea to refined tea, tea shipment from factories to stores, the storage of tea in stores and the time when the tea arrives at your home.

Proper tea storage can ensure the rights and interests of consumers, give full play to the economic value of tea and be of benefit of the tea industry. As customers, our main concern is with tea storage at home — how can tea be stored to remain fresh and flavorful.

茶叶是疏松多孔的干燥食品，保质期较长，但它具有很强的吸湿性，且容易氧化和吸收异味。若茶叶贮存不当，则很容易陈化变质。茶叶陈化变质是茶叶中各种化学成分氧化、降解、转化的结果[1]。

As a kind of dry food (generally, the moisture content of tea leaves should not exceed 6%, the ambient relative humidity should not exceed 50%), tea has a fairly long shelf life. However, its porous structure is apt to absorb moisture and smell from the air around it. That is to say, improper storage can lead to deterioration of tea quality. The aging and deteriorating of tea is the result of the oxidation, degradation and transformation of various chemical components in tea.

（2）影响茶叶品质的环境因素 / External Factors Affecting Tea Stability

在茶叶贮存过程中，对品质影响最大的是温度、湿度、氧气和光线。温度是引发茶叶品质变化的主要因素之一，温度越高，变化越快。湿度会促使茶叶含水量增加，从而加速茶叶的氧化速度甚至造成霉变。氧气会与茶叶中的儿茶素、维生素 C 等营养物质发生氧化反应，加速茶的陈化。光线照射也会加速茶叶内含物质的化学反应，在降低叶绿

---

① 茶叶陈化变质主要包括叶绿素的降低、茶多酚的氧化聚合、维生素 C 的减少、酯类物质的水解、胡萝卜素的氧化、氨基酸的变化以及香气成分的变化。

素的同时产生具有陈味特征、日晒味特征的物质。此外，茶极易吸附异味的特点要求贮存空间避免出现味道浓烈的物品或食品。因此，各种茶的贮存方法虽有不同，总的贮存原则包括干燥（包括茶自身干燥和贮存环境干燥）、低温（部分茶类）、阴凉（避光）、无异味[①]。

A number of factors may influence the storage of tea, and among them temperature, humidity, oxygen and illumination are the most important ones. In term of temperature, high temperature is a major cause of deterioration of tea. Moisture could accelerate the oxidation rate of tea and even cause tea to go moldy. Oxygen will react with nutrients in tea which may also lead to the aging of tea. Illumination will accelerate the chemical reaction which produces substances with aging and stale smell. In addition, considering the fact that tea is easy to absorb smell, it should be kept away from anything with strong smell. Although the storage methods of different teas are slightly different, the general principal rules are the same: stored in sealed bags in such a way that tea is not affected by humidity, temperature and smell.

（3）不同茶类的贮存原则 / Tips for the Storage of Different Teas

就绿茶和黄茶而言，最方便的贮存办法是将茶叶放入袋中并置于

---

① 《中华人民共和国国家标准：茶叶贮存》（GB/T 30375—2013）对不同茶类贮存的温度与湿度有如下规定：绿茶贮存宜控制温度 10℃以下、相对湿度 50%以下；红茶贮存宜控制温度 25℃以下、相对湿度 50%以下；乌龙茶贮存宜控制温度 25℃以下、相对湿度 50%以下，对于文火烘干的乌龙茶贮存，宜控制温度 10℃以下；黄茶贮存宜控制温度 10℃以下、相对湿度 50%以下；白茶贮存宜控制温度 25℃以下、相对湿度 50%以下；花茶贮存宜控制温度 25℃以下、相对湿度 50%以下；黑茶贮存宜控制温度 25℃以下、相对湿度 70%以下；紧压茶贮存宜控制温度 25℃以下、相对湿度 70%以下。

冰箱冷藏①，但需要注意的是，通常家庭冰箱都会放置一些其他食物，而茶叶又容易吸味，所以在贮存的时候，务必把茶叶密封包裹好。绿茶要趁新鲜喝，一般最佳赏味期在 18 个月左右。

The best way to store green tea and yellow tea is to put them into sealed bags and store them in the refrigerator. It should be noted the refrigerator is usually crammed with various food. As tea is easy to absorb smell, be sure to wrap it tightly. Generally, the shelf life of green tea is about 18 months.

锡罐储茶

一般而言，红茶放在阴凉避光、干燥、无异味的条件下即可，高档红茶可以密封后放入冰箱②。但红茶不宜久放，一般最佳赏味期在两年左右。

Black tea can be stored at room temperature and it should be stored in dry, odor-free places, avoiding excessive heat and direct sunlight. High-quality black tea can be kept sealed and refrigerated. Generally, the shelf life of black tea is about two years.

通常情况下，乌龙茶的保质期与其发酵、焙火程度有关。乌龙茶

---

① 关于名优绿茶的贮藏和保鲜，农业农村部曾刊文 "低温（5℃以下）、低湿、避光、阻氧有利于绿茶品质的保鲜。保鲜库、冰箱和冷柜能够实现低温低湿和避光的目的，目前已在生产企业和销售终端中普遍应用。"

② 国外知名茶叶品牌对于茶叶家庭贮存的建议是存放在室温下即可。这或许是因为在欧美家庭以饮用红茶袋泡茶为主，消耗较快，且冰箱食物较多，容易串味。英国家庭通常将茶叶置于厨房的橱柜中。

应贮存在通风、干燥、清洁、阴凉、无阳光直接照射的地方。保质期由生产者根据产品的类型、包装材料和贮存条件等因素自行确定。一般而言，发酵较轻、焙火较轻的乌龙茶，品质接近绿茶，相对保质期较短，建议密封后放入冰箱；重发酵的乌龙茶则可在合适的条件下长期存放。

Oolong tea should be stored in a place which is well-ventilated, dry, clean, avoiding excessive heat and direct sunlight. Generally, the shelf life of oolong tea is related to the degree of its fermentation—the lighter the fermentation, the shorter its shelf life. Lightly fermented oolong tea is similar to green tea in nature; therefore it should be sealed tightly in bags and stored in the refrigerator, while heavily fermented oolong tea can be kept at room temperature for quite a long time. As oolong tea is a complex and multi-faceted tea, the shelf life is determined by the producer based on factors such as product type, packaging materials, and storage conditions.

黑茶作为发酵后茶，在发酵的过程中具有越陈越香的特点，即存放的时间越久，口感就越醇厚。这一特性让黑茶不仅在消费属性上具有很大潜力，在投资收藏方面也具有很高的经济属性。黑茶虽然可以长期存放，但也是有适宜期限的[1]。黑茶存放要求阴凉忌日晒、通风忌密闭、洁净忌异味、干燥忌潮湿。

陶罐储茶

---

[1] 一般认为砖茶、千两茶等可存放10~15年，散茶5~10年为佳。在此期限内，茶品的经济、品饮价值较高。

如果打算长期存放成件的黑茶，最好不要拆开原包装。当然用陶罐贮存也是不错的选择。

As a post-fermented tea, dark tea tastes better as it ferments with time. That is to say, the longer it is stored, the better the taste is. Therefore, dark tea is not only for consumption but also for collection. Although there is no exact shelf life for dark tea, it does have a "best-before date". The storage of dark tea requires an environment which is cool, clean, dry, well-ventilated and odor-free. We may either remain the tea package intact or put it in pottery jars.

对消费者而言，茶叶存储得好，可以让他们在家长期有好茶喝；对企业而言，好的仓储条件能促使黑茶的风味品质得到改善，从而创造出更高的经济效益。正因如此，近几十年来，黑茶尤其是普洱茶的贮藏成为学术界及业界研究开发的热点。

If dark tea is properly stored, customers may enjoy better tea at home for a long time; meanwhile, tea companies may reap more benefits since good storage promotes the price as well as flavor of tea. Because of this, the storage of dark tea, especially that of Pu'er tea, has long remained a hot topic in the tea industry as well as the academic research.

白茶可以常温放在干燥、无异味的地方，且需密封存放。白茶有一定的收藏价值，它与黑茶一样，贮存年份越久，茶味越醇厚，故有"一年茶、三年药、七年宝"一说。与黑茶贮存不同的是，白茶需要密封存放。虽然时间可以让白茶自然增值，但前提是茶叶品质要好且保存得当。

White tea should be placed in a dry and odor-free place at room temperature. It should be noted that white tea needs to be sealed tightly. It is believed that white tea becomes more flavorful with storage just like dark tea. As an old saying goes, "first year tea, third year medicine and seventh year treasure", the value of white tea increases with time. Of course, it is achieved on the premise that the tea is of good quality and it is properly preserved.

紫砂罐储茶（视觉中国　供图）

花茶一般不放入冰箱，因为低温会降低花茶的香味，常温存放于阴凉、干燥、洁净处即可。相对而言，花茶的保质期较短，通常在一到两年不等，且开封后的花茶香气易散发，故对花茶应趁新鲜及时品饮。

Do not put scented tea in the refrigerator because low temperature will impair its aroma. Store it in a cool, dry and clean place at room temperature. Generally, the shelf life of scented tea ranges from 1 to 2 years. Once opened, it will soon lose its flavor.

6.2 茶叶贮存方式的变迁 / Evolution of Storage of Tea

自古以来，人们对茶的贮存就十分讲究，陆羽在《茶经》中对茶的贮存器具和方法[1]曾有提及。在千百年的漫长岁月中，民间常用瓦罐石灰存茶。近几十年来，茶叶的贮存发生了很大的变化。科技的发展，尤其是冰箱的出现让茶叶贮存变得方便而有效。当下，常用的茶叶贮存方法有瓦罐石灰法、真空贮存法、抽气充氮法及最便捷最普及的低温贮存法。

Since ancient times, people have attached great importance to the storage of tea. In *The Classic of Tea,* Lu Yu illustrated the storage of tea. For quite a long time, tea was kept in earthen jars with quick lime. In recent decades, the packaging and storage of tea has changed greatly. The commonly used methods for storing tea include storing tea in the refrigerator or earthen jars with quick lime, vacuum storage, and nitrogen-filling method. Among them, storing tea in the refrigerator is the most convenient and easiest way.

以碧螺春茶的贮存方式变迁为例，传统碧螺春茶采用瓦罐石灰法贮存，人们用牛皮纸把茶叶包好，石灰用纱布装袋，将包好的茶叶分层环排于瓦罐的四周，然后在茶叶中间放置石灰袋，放好后密封瓦罐[2]。瓦罐放于阴凉、避光处，罐内石灰一个月左右更换一次。如此贮

---

① 《茶经·二之具》中的"育"就是茶的存储用具。"育，以木制之，以竹编之，以纸糊之，中有隔，上有覆，下有床，傍有门，掩一扇，中置一器，贮煻煨火，令熅熅然，江南梅雨时焚之以火。"吴觉农在《茶经述评》中写道："育，既是一种成品茶的复焙工具，也是一种封藏工具。用微弱无焰的火烘茶，这是一种低温长烘用以防潮的方法。"

② 比较讲究的做法，还包括事先将瓦罐用火烤干及选用新鲜石灰等。茶人三部曲之《南方有嘉木》中有相关描述。

存，茶叶可保鲜半年以上。今天，我们只需将碧螺春茶密封后放入冰箱，茶叶便可保鲜至来年新茶上市。

Take Biluochun tea as an example. The storage of Biluochun tea was complicated in the old days. People used kraft paper to pack tea and then put it in a sealed earthen jar with quick lime. The jar was put in a place avoiding direct sunlight and excessive heat and people replaced the lime in the jar once a month. In this way, tea could be kept fresh for at least half a year. Today, thanks to the refrigerator, storing Biluochun tea is much easier. Tea is first sealed tightly, and then put into the refrigerator at the desired temperature. In this way, Biluochun tea can be kept fresh for a year.

虽然茶的贮存手段今非昔比，茶本质上是一种食物，所以除特殊茶类外，大家还应趁新鲜品饮。此外，生活中，我们最好少量购买茶叶。如果一次性购买了较多茶叶，建议将大部分茶叶存放于密闭罐中，留出少量茶叶放在另一个罐中供日常饮用。如此将减少茶叶暴露在光照下和空气中的时间。如果茶叶变质或走味了，就不要再喝，以免对身体造成危害。

Tea is a kind of food in nature, so it should be consumed while it is fresh. Of course, there are some exceptions such as dark tea and white tea. It is kindly suggested that as consumers, we buy small amount of tea. If we have purchased a lot of tea, we can put most of the tea in a large sealed container, and leave a small amount of tea to another pot for everyday use. In this way, tea can be better stored by avoiding being exposed to light and air every time you drink tea. For the sake of health, do not drink the tea which has lost its flavor.

小贴士：

1. 每年六七月，江南地区都会出现持续雨天的梅雨季，为防止这个时节茶叶受潮变质，可采用以下三个方法。一是分类密封，即鲜嫩的绿茶、台湾高山乌龙和轻焙火的铁观音一类的茶置于冰箱，其他茶类各自密封，尽量不混放；二是离地离墙，即因梅雨季节，湿气大，地面和墙体会积聚大量水汽，故茶叶最好置于架子上，与墙面、地面保持 20 厘米左右的距离；三是控制好家中的湿度，尤其是下雨时，要关闭家中门窗，隔绝窗外湿气。可使用除湿器，没有除湿器的话，使用空调除湿功能。

2.《茶叶贮存》国家标准（GB/T 30375—2013）规定了各类茶叶产品贮存的要求、管理、保质措施、试验方法。但该标准并未列出各类茶叶可以贮存的时间。结合不同茶类的国家标准，绿茶、黄茶、红茶均表述为"应符合 GB/T 30375 的规定"。乌龙茶的保质期如文中所说，多样且灵活。《白茶》国家标准（GB/T 22291—2017）写明白茶的贮存"应符合 GB/T 30375 的规定。产品可长期保存。"《地理标志产品：普洱茶》国家标准（GB/T 22111—2008）写明"在符合本标准的贮存条件下，普洱茶适宜长期保存。"综上，在国家标准中，写明可长期保存的只有白茶和黑茶。

07 茶之效
Health Properties of Tea

在世界三大非酒精类饮料——茶、咖啡和可可中，茶的消费人群最大，达 20 多亿。作为日常饮品的茶，香气迷人、口感宜人，既可以解渴补水，又可放松身心；而作为健康饮品的茶，保健功效亦为世人所赞。

Among the three major non-alcoholic beverages in the world, tea, with more than 2 billion consumers, is much more popular than its counterparts coffee and cocoa. As a beverage, tea quenches thirst and helps people relax with its pleasant aroma and taste; as a health promoter, tea is long believed to boost health and strengthen the immune system.

数千年前，我们的祖先就发现茶的药用和保健功效。历代古籍如《神农本草经》《本草纲目》等对茶的功效都有翔实记载。如《神农本草经》中"神农尝百草，日遇七十二毒，得茶而解之"为对茶叶药用的最早记录，突出的是茶的解毒功效。东汉、三国（25—220）时期的医学家华佗在《食论》中提出了"苦茶久食，益意思"，这是对茶药理功能的最早记述。陆羽在《茶经》中写道："茶之为用，味至寒。为饮，最宜精行俭德之人，若热渴、凝闷、脑疼、目涩、四肢烦、百节不舒，聊四五啜，与醍醐、甘露抗衡也"。唐代著名医学家陈藏器在《本草拾遗》中写道"诸药为各病之药，茶为万病之药"。此话虽是夸大之词，但也说明茶有多方面的疗效。李时珍在《本草纲目》中对茶的药效及附方等进行了详尽的记录与阐释。近年来，我国学者基于古代文献总结出茶的二十多项传统功效 ①。

Chinese ancestors discovered the curative powers of tea thousands of years ago. Many Chinese ancient books had detailed records on its health properties. For example, the discovery of tea was closely linked to the detoxification effect of tea, as recorded in *Shennong Bencao Jing*[2], the most well-known work mainly attributed to Shennong; Lu Yu, the tea sage, also listed a number of benefits of tea in the famous *The Classic of Tea*; Chen Cangqi, a famous medical expert in the Tang Dynasty highly praised tea as a magic panacea, which might be a slight exaggeration but it revealed various curative powers of tea. In *Bencao Gangmu*[3], the most famous and complete medical book in the history of traditional Chinese medicine, Li Shizhen, a

---

① 茶的传统功效包括：解毒、少睡、安神、明目、清头目、止渴生津、清热、消暑、消食、醒酒、去肥腻、下气、利水、通便、治痢、去痰、祛风解表、坚齿、治心痛、疗疮治瘘、疗饥、益气力、延年益寿及其他。

② *The Divine Farmer's Herb-Root Classic* 和 *Shennong's Classic of Material Medica* 为《神农本草经》的另两种广为人知的译法。

③ *The Compendium of Materia Medica* 为《本草纲目》另一种广为人知的译法；中国日报（China Daily）将其译为 *Herbal Foundation Compendium*。With almost 2 million Chinese characters, the book lists 1,892 kinds of medical substances. Besides Chinese herbal medicines, it includes animals and minerals used as medicinal substances. It is so famous that it has been translated into more than 20 languages. Even today it is often used as a reference book.

famous medical scientist in the Ming Dynasty, made detailed records of the use of tea either alone or in combination with other herbs for medical uses. In recent years, Chinese scholars have summarized over 20 health benefits of tea from dozens of ancient texts.

　　茶最初在日本的传播也与其保健功效息息相关。有"日本茶经"之誉的《吃茶养生记》主要论述的就是茶的药物性能。茶最初传入欧洲，就是通过其药用价值吸引顾客。不久，当茶传入英国时，也被当作一种健康的药用饮料销售。历史上，1658年出现的第一则茶叶广告主要宣传的也是它的药效，且是"得到所有医师认可的"[①]。这当然不仅仅是一个促销噱头，因为茶确实具有神奇的保健功效。

The health proprieties of tea also contributed to the popularity of tea in Japan. *Drinking tea for Nourishing Life*, reputed as "*Japanese Classics on Tea*", dwelled on the medicinal properties of tea. When tea was introduced into Europe, it was the medicinal value that fascinated customers. Later when it was introduced to England, tea was marketed as a healthy, medicinal drink. The first advertisement of tea in history appeared in 1658 stressed the excellent health properties of tea which was "approved by all physicians". This was not just a promotional gimmick. Tea does contain marvelous medicinal values.

---

　　① 　这则刊载于报纸上的广告原文如下：That Excellent, and by all physicians approved, China Drink, called by the Chinese Tcha, by other nations Tay alias Tee, is sold at the Sultaness-head, a coffee shop in Sweetings Rents by the Royal Exchange, London.（得到所有医师认可、品质优异的中国饮料，中国人称为"Tcha"，其他国家称为"Tay"或者"Tee"，现在伦敦皇家交易所旁 Sweetings Rents 的 Sultaness-head 咖啡店出售。）

茶园春意（静观 供图）

20世纪80年代，医药界倡导回归大自然，很多国家的科研人员尝试从植物等天然物质中提取有效成分研制药物和保健药品。很多科学家对茶叶与人体健康的关系进行研究①。现代医学研究认为，茶的保健功效主要包括提高免疫力、清胃消食、坚齿防龋、消脂瘦身、明目利尿、兴奋提神等②。

In the 1980s, the pharmaceutical industry advocated "back to mother nature". Since then, there has been a worldwide trend on the research of herbal medicines and/or phyto-pharmaceuticals, with tea being one of the research focuses. Recent studies indicate that tea is beneficial to health by boosting immune system, promoting digestion, strengthening teeth and preventing caries, losing weight, improving eyesight, having a mild diuretic effect, relieving fatigue and delighting the soul, etc.

---

① 目前，每年与茶相关的期刊文献约为 14 000 篇，其中约有 50% 的研究是从医学角度入手。

② 茶的保健功能还包括：改善血液组成、降血脂、预防心血管疾病、降血压、降血糖、消炎灭菌、醒酒解毒、抗过敏、抗焦虑及对神经退化性疾病的预防、抗癌抗突变等。

7.1 茶叶中的健康因子 / Constituents in Tea

随着茶叶化学和药理学研究的深入，人们不断从茶叶中发现有利于人类健康的功能性成分，如茶多酚、咖啡因、茶氨酸、茶多糖、茶黄素、矿物元素、维生素等。

Studies in tea chemistry and pharmacology have discovered many functional components beneficial to human health, such as tea polyphenols, caffeine, theanine, tea polysaccharides, theaflavins, mineral elements and vitamins.

（1）茶多酚 / Polyphenols

茶多酚[①]是茶叶中最主要的功能性成分。鲜叶中茶多酚总量一般占干物质重量的14%~33%[②]。儿茶素是组成茶多酚的主要物质，也是最重要的保健因子。现代科学研究表明，茶多酚具有多种生理活性和保健功能，如抗氧化、清除自由基、抗辐射、抗菌、抗病毒、降血脂、降血糖等。作为一种天然的、功效突出的抗氧化剂，茶多酚被广泛应用于保健食品中。茶多酚保健品主要具有降血脂、提高免疫力和减肥等保健功效。

Tea is rich in compounds called polyphenols. The total polyphenols in fresh tea leaves range from 14% to 33% (dry weight basis). Among them, Catechins, the main dominant group, are the most important health component. Modern scientific research shows that tea polyphenols have a variety of health

---

[①] 茶多酚是茶叶中儿茶素类、丙酮类、酚酸类和花色素类化合物的总称，是形成茶叶色香味的主要成分之一。

[②] 数据来源于《中国茶经》（修订版），不同书籍给出的数据存在细微差异。

properties, such as anti-oxidation, free radicals scavenging, anti-radiation, anti-bacteria, anti-virus, lowering blood lipid and blood glucose, etc. As a natural and effective antioxidant, tea polyphenols are widely used in health food. The main functions of such products include lowering blood lipid, boosting the immunity system as well as weight control.

（2）咖啡因 / Caffeine

咖啡因又名咖啡碱[①]，1927年首次在茶叶中发现。咖啡因是茶叶中含量最高的生物碱，一般占茶叶干物质的 2%~5%。一般而言，大叶种茶树的咖啡因含量相对较高，细嫩茶叶较粗老茶叶含量高。因能兴奋中枢神经系统，咖啡因被广泛应用于抗疲劳、抗焦虑。近年来，国内外大量研究表明咖啡因对癌症及肿瘤有一定的防治作用；此外，咖啡因对于阿尔茨海默病、帕金森病等均有一定的预防及治疗效果。

In 1927, caffeine was discovered in tea. It is the major alkaloid present in tea, with a general content ranging from 2% to 5% (dry weight basis). Generally speaking, big-leaf tea contains more caffeine than small-leaf one; fine and tender tea contains more than coarse one. Caffeine has been widely used in relieving fatigue and anxiety because it is very effective in exciting and stimulating the central nervous system. In recent years, studies have shown that caffeine has a great potential for the treatment of cancers and tumors; in addition, caffeine has a positive effect on the prevention and treatment of Alzheimer's disease and Parkinson's disease.

研究发现，咖啡因的析出量在很大程度上与冲泡方式有关，即温

[①]　茶叶中的生物碱包括咖啡碱、可可碱、茶碱，其中含量最多的是咖啡碱。

度越高、茶叶浸泡时间越长，咖啡因析出越多。需要指出的是，虽然咖啡因的生物活性被广泛利用，但如何衡量咖啡因对人体的利弊还是个问题，且摄入过量的咖啡因会对人体产生短期的负面作用[①]，故如何降低茶叶中咖啡因的含量[②]也是近些年研究的方向之一。

Studies show, to a large extent, the amount of caffeine a cup of tea contains is largely determined by how it's brewed; hotter water and a longer steeping time lead to more caffeine being released. Although the biological activity of caffeine is widely used, whether caffeine is a miracle drug or junk is hard to say. A problem some people suffer is caffeine sensitivity[③]. Luckily, decaffeinated tea has been successfully produced.

（3）茶氨酸 / Theanine

茶氨酸是茶叶的特征性氨基酸，也是茶叶中含量最多的游离氨基酸。1950 年由日本学者首次从玉露茶中分离得到并命名。茶鲜叶中茶氨酸的含量一般在 1%~2%，在某些名优茶中含量超过 4%。茶氨酸是茶汤鲜甜味的主要成分，与绿茶品质密切相关。一般来说，茶叶嫩度越高、级别越高，其茶氨酸的含量也越高[④]。茶氨酸的保健作用主要在于松弛神经[⑤]。此外，茶氨酸能够缓解咖啡因的刺激效果，延缓咖啡因

---

① 健康成年人咖啡因安全摄入建议值是每天不超过 400 毫克，相当于 3~4 杯冲泡完全的茶。超过 400 毫克，可能会导致焦虑、紧张、头晕和烦躁等症状。咖啡因摄取量与个体咖啡因敏感度及分解代谢速度有关。

② 国际知名茶品牌如立顿、川宁等，均有去咖啡因茶。

③ The maximum safe daily caffeine intake is 400mg for most healthy adults. Any more than 400mg of caffeine per day is likely to make people feel confused, nervous, dizzy, and irritated.

④ 因光合反应会减少茶氨酸的含量，日本人用遮光的方式，种植高级绿茶玉露以提高茶氨酸含量，而玉露的甘甜鲜醇便来自茶氨酸。

⑤ 茶氨酸能增加多巴胺的分泌，多巴胺可以有效降低焦虑感和忧郁感，因而茶氨酸能让人安神镇静。

的吸收。

Theanine is a special amino acid in tea. It was first extracted from gyokuro, a high-grade tea made in Japan from the leaves of shaded bushes, by Japanese scholars in 1950. As a distinctive element in tea, it generally constitutes 1%–2% in fresh tea leaves (dry weight basis), and more than 4% in some high-quality teas. It contributes to the fresh and sweet taste of the green tea. Generally speaking, tea with superior quality contains more theanine than the inferior[1]. In terms of health properties, theanine helps people relax; besides it alleviates the stimulating effect of caffeine, which makes tea a better choice in delighting the soul and strengthening the will than coffee.

（4）茶多糖 / Tea Polysaccharides

茶多糖与原料的嫩度和茶类有关。一般说来，茶多糖的含量随着原料粗老程度的增加而增加，黑茶、乌龙茶中茶多糖的含量高于红茶和绿茶。茶多糖有防辐射、抗凝血、抗血栓、降血糖和增强机体免疫力等多种生理功能。近些年，研究发现茶多糖在糖尿病的治疗上功效显著，此外茶多糖还被广泛应用于保健品、化妆品和医药等领域。

Tea polysaccharides are related to the tenderness of tea, which means the tenderer the raw material, the lower tea polysaccharides content. Besides, different types of tea have different polysaccharides content. Generally, dark tea and oolong tea contain more tea polysaccharides than black tea and green tea. Tea polysaccharides are known for functions of anti-radiation, anti-

---

[1]　Since photosynthetic reaction will decrease theanine, the Japanese shade the tea bush to discourage photosynthetic reaction. In this way, the amount of theanine in the famous tea gyokuro is greatly increased, which results in the distinctive sweet and fresh taste of gyokuro.

coagulation, anti-thrombosis, lowering blood sugar and enhancing immunity. In recent years, it has been found that tea polysaccharides are effective in the treatment of diabetes, and it has been widely used in many fields such as health food, cosmetics and medicine.

7.2 茶的保健功效 / Health Benefits of Tea

　　古往今来，有关茶的保健功效和健康疗效的记录和研究非常丰富，有些功效是作为普通消费者的我们日常能够感受到的，也有一些是我们不太熟悉的。

Tea is crammed with a variety of health properties. Some are familiar to common tea consumers like you and me, and other benefits of drinking tea are less known to us.

（1）生津止渴、清热解暑 / Quench the Thirst and Beat the Heat

　　茶作为一种古老的饮品，其最质朴的功效就是生津止渴，是茶多酚、咖啡因、多种芳香物质和维生素等成分综合作用的结果。在生津止渴的同时，茶以其丰富的口感愉悦了饮茶人。

As an ancient beverage, tea quenches the thirst better than water, which is the result of constituents such as polyphenols, caffeine, aromatic substances and vitamins. Tea is not only a great thirst quencher, but it offers delicate and pleasant flavors.

解暑绿茶（马娜　供图）

炎炎夏日，酷暑难当。一杯绿茶最是清热解暑；新白茶消暑效果也不错。当下，冷泡的乌龙茶（尤其是葡萄乌龙、白桃乌龙等加味乌龙茶）也是很多年轻人解暑的新选择。

On hot summer days, a cup of green tea or new white tea[①] is the best way to beat the heat. Besides, oolong tea (mostly flavored oolong tea, such as grape flavored oolong tea, white peach flavored oolong tea, etc.) brewed in cold water, is very popular among young people.

（2）提神醒脑、消除疲劳 / Relieve Fatigue and Delight the Soul

说到提神醒脑，年轻人首先想到的是咖啡及其功效物质咖啡因。众所周知，咖啡因最显著的功效之一就是兴奋中枢神经系统。与咖啡一样，茶叶也有提神醒脑的功效，因为茶叶同样含有咖啡因。虽然在茶叶中发现咖啡因尚不足百年，但茶的提神醒脑这一功效数千年前就为古人所识、所用[②]。此外，由于茶叶所含的儿茶素、茶氨酸能改变咖啡因的吸收速度，延长咖啡因在体内的停留时间，故相较于咖啡，茶的提神醒脑功效更温和、更持久。

When it comes to getting mentally refreshed, young people will immediately think of coffee and its potent substance—caffeine, an substance which can excite the central nervous system. However, Chinese people have discovered the same effect in tea since ancient times. Modern research attributes this effect to caffeine that tea contains. Compared with coffee, the refreshing effect of tea is milder and longer-lasting because the catechin and

---

① 新白茶译为 new white tea; 老白茶则译为 aged white tea.

② 传说达摩祖师在少林寺面壁，誓言无眠禅定九年以警醒世人。后因瞌睡，达摩把眼皮撕下丢在地上，结果长出一棵树。采了树叶煮水饮用，提神醒脑。禅寺饮茶之风就此开启。

theanine in tea can prolong the effect of caffeine in the body.

茶除了可以提神醒脑，还能够消除疲劳。研究表明，茶氨酸可以明显促进大脑多巴胺释放，提高其生理活性。多巴胺活性的提高，有助于舒缓精神，使人感到放松和平静，从而达到抗疲劳的作用。

In addition, tea can relieve fatigue and delight the soul. Studies have shown that theanine can significantly increase dopamine in the brain and improve its physiological activity which makes people feel relaxed and calm.

提神醒脑、消除疲劳以绿茶、新白茶和普洱生茶效果最佳。

For getting refreshed and relieving fatigue, green tea, new white tea and raw Pu'er tea are the best choices.

（3）清胃消食、帮助消化 / Promote Digestion

茶叶中芳香油、生物碱等物质可以刺激胃液分泌、加强肠胃蠕动，有良好的助消化效果。这一功效很早就为中国人所知，在不少书籍中都有提及[①]。这一功效对于边疆地区的人们非常重要，有助于他们调节饮食结构。如在西藏等地，人们以肉奶类为主食，饮食中含有大量的脂肪和蛋白质。茶可以帮助消化，解除油腻，并补充矿物质和维生素的不足，故这些地区的人们"不可一日无茶"。

Some ingredients such as aromatic oil, alkaloids of tea are helpful

---

① 如在《红楼梦》中，就有吃一种名为"女儿茶"的普洱茶以助消化的片段。《红楼梦》第 63 回"寿怡红群芳开夜宴 死金丹独艳理亲丧"中，一行人来怡红院查上夜，正逢宝玉吃了面怕积食还没睡，林之孝家的便向袭人等笑说："该沏些个普洱茶吃。"袭人、晴雯回道："沏了一盅子女儿茶，已经吃过两碗了。"

for digestion, which has long been recognized and widely recorded by Chinese people. This function is of greater importance for people living in frontier areas. In Xizang for example, the local people habitually take beef and milk as their staple food. Lack of vegetables and fruits may lead to the deficiency of minerals and vitamins. Fortunately, tea helps balance their diet by providing vitamins. Therefore, for the locals, tea is a necessity for daily life.

六大茶类均可助消化，其中又以乌龙茶、普洱茶的去油解腻功效最为突出。

Although all types of teas are good for digestion, oolong tea and Pu'er tea stand out as being more effective than the rest.

（4）清新口气、预防龋齿 / Improve Dental Health

日常生活中，我们在饮茶尤其是饮用绿茶后，会明显感到口气清新。饭后用茶漱口的习惯由来已久，如苏东坡就曾介绍饭后用茶漱口可固齿[①]；《红楼梦》也曾多次提及饭后的这杯漱口茶[②]。饭后以茶漱口，既可清除口腔内的食物残渣，也可以解油腻、清新口气。此外，饮用

---

① 苏轼《仇池笔记》中有《论茶》一则，其中写道："吾有一法，每食已，以浓茶漱口，烦腻既出，而脾胃不知。"（"I'd like to recommend a daily usage of tea-rinsing the mouth with strong tea. It helps clear away the greasy feeling without upsetting the stomach."）

② 如《红楼梦》第3回"贾雨村黄缘复旧职，林黛玉抛父进京都"中，就描述了饭后用于漱口的茶，详见本书第167页注释①。第28回"蒋玉菡情赠茜香罗，薛宝钗羞笼红麝串"中有这么一段："一时吃过饭，宝玉一则怕贾母乏捆，二则也记挂着林黛玉，忙忙的要漱口。"第54回"史太君破陈腐旧套，王熙凤效戏彩斑衣"中有这么一段："大家随便随意吃了些，用过漱口茶，方散。"

绿茶，还能预防龋齿。早在 20 世纪 80 年代，日本科学家[①]就确证绿茶能有效预防龋齿。茶叶中的茶多酚能够杀死口腔中的龋齿细菌，使龋齿细菌难以黏附于牙齿表面[②]。因此，近年来，茶叶提取物被加入牙膏中，发挥预防龋齿的作用。

Many of us have learned that green tea can help beat bad breath. Green tea has long been regarded as the recipe for fresh breath. Su Dongpo, a great poet of the Song Dynasty, once shared his personal daily practice of rinsing his mouth with tea after each meal, which was believed to strengthen teeth. In the famous novel *A Dream of Red Mansions*, there are a couple of episodes describing gargling with tea after dinner. In addition to beating bad breath, tea also helps prevent tooth decay, which was supported by the research conducted by Japanese scientists in the 1980s. It has been proved that tea polyphenols can kill bacteria in the mouth, making it difficult for dental caries bacteria to adhere to the tooth surface. Therefore, the extract from tea has been added to toothpaste to prevent dental caries.

清新口气、预防龋齿首推口味清香、汤色较淡的绿茶；新白茶的口腔保健效果也不错。日常生活中，我们可在饭后用绿茶或白茶汤漱口 2~3 次，每次 30 秒左右。

Green tea which has light taste and color is effective in oral nursing, so is white tea. In daily life, you may rinse the mouth with green tea or white tea for about 30 seconds for 2 or 3 times after each meal.

[①] 这项 1981 年开始的研究由大阪大学及大洋化学工业公司中心科学实验室主导，确证饭后一杯绿茶能够预防龋齿。

[②] 在中国科协与新华网联合打造的专题栏目"科学前沿大师谈"中，陈宗懋院士介绍了茶的防龋齿功效。

（5）消脂瘦身、温和减肥 / Help Lose Weight

肥胖成了日益严重的全球性健康问题，可诱发高血压、糖尿病等多种疾病，给人们的健康带来巨大威胁。有研究证实茶叶具有良好的降脂减肥效果[①]。研究还发现六大茶类均有一定的防治肥胖、改善相关代谢综合征的作用。需要注意的是，日常生活中如不加节制地吃喝，或者喝茶时加糖加奶，再配上饼干和巧克力等甜食，靠喝茶减肥则是不可能的。

Obesity is a concern because of its implications for the health as it increases the risk of many diseases including hypertension, diabetes etc. The link between consuming tea and the rate of reducing fat has been confirmed. According to the research, all teas have certain effects on beating obesity and improving related metabolic syndrome. However, it is easy to upset the health benefits of tea. If you're someone who always overeats or opts for sugar, milk, biscuits and chocolate with tea, you are unlikely to lose weight.

六大茶类中，绿茶、乌龙茶和普洱茶在控制体重方面功效比较突出。

Among six basic teas, green tea, oolong tea and Pu'er tea stand out as being more effective than the rest in losing weight.

我们应该清楚虽然茶富含多种对人体有益的物质，但它不能代替水果蔬菜。坚持合理饮食，适量运动，形成良好生活习惯，才是健康

---

① 该研究来自浙江大学周继红、王岳飞等。研究表明，一方面，茶叶中的活性成分能够有效抑制脂质的形成与积累；另一方面，茶叶及其提取物还可以调节饮食吸收、促进脂质排泄。此外，茶叶还可以通过调节肠道菌群、调节激素分泌、促进能量消耗等机理改善肥胖及相关代谢综合征。

的保障。虽然茶有多种保健功效，但绝非万能，我们不能夸大其效果，盲目依赖茶疗。

It should be pointed out that although tea is rich in many substances beneficial to heath, it cannot be a substitute for fruits and vegetables. A healthy diet and lifestyle as well as regular exercise are more important to one's health than tea. Besides, never overwhelmingly depend on the health properties of tea. Tea is not a magic cure-all for all diseases.

7.3 科学饮茶 / A Cup of Health

水是人体需要量最大的营养素，人体每天约需要 2000 毫升的水。喝茶不仅是很好的补水方式，口感还比较好。

Tea is a pleasant way to help people achieve the recommended daily fluid intake of about 2 liters a day. The first ingredient in tea is water, which is needed by the body to run at its best.

不同的茶所含成分是有差异的，这些差异不仅体现在茶汤颜色、口感和香气上，也反映在营养元素构成及功效上。有实验表明，在绿茶、乌龙茶、黑茶和红茶中，绿茶的维生素 C 和茶多酚含量最高；红茶的茶多酚含量最低但咖啡因含量却最高；黑茶和乌龙茶的茶多糖含量较高。因此，补充维生素 C 选择绿茶较好[①]；提神醒脑选择红茶，且红茶的茶多酚含量较低，比较适合脾胃虚弱的人群，这便是人们常说"红茶暖胃"的缘由；而乌龙茶和黑茶的降脂减肥效果总体优于红茶和

---

① 100克高级绿茶通常含300~600毫克的维生素C，每天喝2~3杯这样的绿茶基本上就可以满足人体对维生素 C 的需要。

绿茶。

Teas are slightly different in ingredients and functions, which are not only reflected in the color, taste and aroma of tea infusion, but also in nutrients and health properties. It has been proved that among four basic teas commonly consumed, green tea has the highest levels of vitamin C and tea polyphenols; black tea contains the most caffeine but the least tea polyphenol; dark tea and oolong tea have highest levels of tea polysaccharides. Therefore, it is better to choose green tea for supplementing vitamin C; the black tea is more suitable for refreshing the mind, meanwhile it is suitable for people who suffer spleen and stomach problems because black tea is less stimulating; dark tea and oolong tea offer a stronger weight loss solution than others.

健康饮茶不仅体现在选择合适的茶品上，也要求饮茶者根据自己的体质和身体状况，因时而异地选择适合自己的茶饮。只有科学合理地饮茶，才能达到养生保健的目的。

To make good use of the health properties of tea, one needs to choose suitable teas according to his/her physique. Besides, drinking the right tea at the right time is also important. Only by drinking tea scientifically, can people benefit from these magic leaves.

（1）健康饮茶法则 / Tips for Drinking Tea

健康饮茶因人而异 / Different Teas for Different People

饮茶需因人而异主要体现在根据不同的年龄、性别、体质甚至职业等因素，选择适合自己的茶品。

People may choose the right cup of tea according to individual conditions such as age, gender, constitution (or physical condition) and even occupation.

就饮茶年龄而言，很多家长不敢给孩子喝茶，认为茶有刺激性，会伤害孩子的脾胃。事实上，只要控制好茶的浓度和饮用的时间，儿童是可以饮茶的[①]。家长可以让孩子适量地饮用温热的淡茶，但要避免让孩子睡前饮茶。

Many parents think children are too delicate to drink tea, which is partly true. Strong tea is harmful to children's spleen and stomach, but a small amount of light tea is helpful. When offering tea to children, make sure it is of suitable temperature. Besides, do not offer tea to children late at night.

老年人的身体调节能力下降，睡眠质量相对不高，故对于茶有些老年人也会纠结于"饮"与"不饮"。我们知道，茶叶中的茶多酚、咖啡因、茶多糖、氨基酸等微量元素对调节血压、血糖、血脂都具有一定的功效，所以老年人可以根据自己身体状况适量饮茶。但患有骨质疏松等疾病的老年人一般应控制饮茶的量或遵循专业医师的建议。另外，不少老年人多年饮茶，口味较重，随着年纪的增长，在饮茶的量和泡茶的浓度上也应有所控制。

For the elderly people, drinking tea seems like a "to be or not to be" question, since tea often leads to a poor sleep. As we know, tea is rich in substances beneficial to human body, so the elderly can choose appropriate tea according to their own physical conditions. However, those suffering from diseases such as osteoporosis should avoid drinking too much tea, or

---

[①] 儿童适当饮用淡茶，并非指冲泡数次，待茶汤寡淡后，再给儿童饮用。此时的茶汤，口感不佳，营养成分含量少。家长可用少量茶叶冲泡，待茶汤温度适宜后，给儿童饮用。

they should follow doctors' instructions. It has been noted that some elderly people who have been drinking tea for many years are "heavy drinkers". Even though, for the sake of health, they should avoid drinking too much strong tea.

就饮茶的性别差异而言，最显著的莫过于女性对花茶的偏好，这主要是因为以花入茶不但外观漂亮、香气怡人、口感也更加温和。此外，冲泡花茶的过程也非常怡人。花茶的保健作用在于鲜花中的芳香油具有镇静调理神经之效。近年来，不少时尚女性喜欢自制花茶。她们选择不同的香花与茶叶搭配，调制出不同风格的花茶。除了花茶外，女性还特别青睐口感微甜的果茶。

The gender difference in tea drinking is significantly reflected in women's preference for scented tea. This is mainly due to the fact that scented tea is more attractive in appearance, more fragrant in aroma, milder in taste. Watching the lovely flowers swelling up in the glass is an enjoyable process. Besides, it is effective in helping settle the nerves. In recent years, many fashionable women take to making scented tea by themselves. Matching different flowers with different teas, they are delighted to make individual scented teas. The fruit tea is another popular tea among women which is very flavorful with a light sweet taste.

就饮茶与体质的关系而言，中医把人的体质分成平和体质、湿热体质等九大类，每种体质各有特点，有专家建议日常饮茶应当依据自己的体质选择适合的茶品。但大多数人并不清楚自己的体质，只是大约知道寒热之别。总的说来，平时畏寒的人群选择红茶为好，因为红茶茶性温和，喝了有祛寒暖胃之感；若平时怕热，则以绿茶为上，因为绿茶性寒，喝了有清凉之感。

According to traditional Chinese medicine, people's constitution is divided into nine categories. Therefore, different teas should be chosen in line with one's constitution. However, most people don't understand much about the concept of constitution, and they may only know a bit of the distinction of physically cold and hot. Despite everything, we should keep in mind that weak people (who are very sensitive to cold and always have cold hands and feet) may choose black tea because it is mild in nature and it warms one's stomach. As for those who are sensitive to heat, it is suggested that they take green tea which will make them feel cool.

就职业而言，平时长时间面对电脑的人群，如白领、教师、科研人员等，可饮用有显著的抗辐射功效的绿茶。建议这一类人群选用较小的茶杯，一来可保证茶汤的口感，二来他们往往长时间坐着工作，使用小茶杯可迫使其简单活动一下——离开座位添茶加水。

In terms of occupation, green tea is a good choice for office workers such as white-collar workers, teachers and researchers who work in the closed environment with radiation contamination. Green tea has a significant anti-radiation effect. Considering their sedentary jobs, we recommend small tea cups (or mugs) for them. On the one hand, small cups ensure fresh taste of tea; on the other hand, they have to go to the tea room from time to time—an easy way of doing exercise in the office.

此外，孕妇及哺乳期女性、神经衰弱人群，以及患有缺铁性贫血[①]或胃溃疡的人群，也要根据医嘱控制饮茶的量和浓度。而睡眠不好或是对咖啡因过敏的人群，则可选用去咖啡因茶。

---

① 李时珍在《本草纲目》中提及"虚寒及血弱之人，饮之既久，则脾胃恶寒，元气暗损。"

In addition, if you are pregnant, nursing, taking any medications or have any medical condition (especially neurasthenia, iron deficiency anemia or gastric ulcer), you had better consult your doctor first. People who suffer from insomnia or who are allergic or sensitive to caffeine may choose the decaffeinated tea which is regular tea that has been processed to removes the caffeine.

健康饮茶因时而变 / Drinking Different Teas at Different Times

因时而异主要体现在两个层面，一是根据季节选择不同的茶[1]；二是根据一天不同的时段选择不同的茶。

Drinking different teas at different times lies in two aspects: choosing different teas at different times of a day as well as in different seasons.

茶叶有寒、热、温、凉性味功能之异，根据四季的时序更迭，选择不同的茶，茶的保健功效事半功倍。春季养生以养肝、护肝为先，故此时可选择气味芬芳的花茶，以振奋精神，散发体内淤积的寒气，促进人体阳气之升发。春季还是品饮绿茶的好时节，一杯鲜爽的绿茶，满口春天的味道。夏季骄阳似火，此时品茗以绿茶为宜。绿茶性寒味苦，可消暑解热、止渴生津；且绿茶富含氨基酸、维生素、矿物质等，所以夏季饮用绿茶，还可以补充营养。秋天干燥，乌龙茶能够生津润肺，这个季节饮用最是合适。冬季天气日渐寒冷，大地冻结，此时适合品饮熟普、红茶之类温和的发酵茶。

Chinese civilization believes different teas are of different properties,

---

[1] 著名茶人静清和写道："健康的喝茶方式，一定要遵循季节的变化。人体从冬至开始，阳气渐次向外浮越，内寒而外热。从夏至开始，阳气由外至内渐渐收敛，逐步内热而外寒，人体这种随节气的微小而玄妙的变化，启示我们健康喝茶要因人而异、因时而变，通过不同茶饮的寒温变化，调整身体的阴阳平衡。"

for example, green tea is thought to be cold while black tea and dark tea are warm, and the rest are somewhere in between. Therefore, drinking the right tea in different seasons may double the benefits of tea. According to traditional Chinese medical science, spring is the best season to nourish the liver. Therefore, scented tea with fragrant smell is the best choice in this season, which may stimulate the spirit, emit the cold in the body, and promote the circulation

桂花乌龙茶（马娜　供图）

of Qi (vital energy). Spring is also the season for green tea, which provides people with the taste of spring. On scorching summer days, green tea is the best choice. It is bitter and cold in nature, which can relieve summer heat and quench thirst; besides, green tea is rich in nutritious elements such as amino acids, vitamins, minerals, etc. Autumn is dry and it is a season to nourish the lung. Oolong tea is the most suitable tea in this season. On freezing cold days, some fermented teas such as ripened Pu'er tea and black tea may warm you better than anything else.

　　一些讲究的茶人甚至会根据节气选择不同的茶品[①]，其中既有文化方面的考量，更有养生方面的考量。

　　Some tea professionals even drink different teas in different solar

---

　　① 古人将一个太阳年划分为季、节、气、候。1年有4季、12节并12气、72候。以春季的立春这一节气为例，立春之后，人体内阳气萌发，而茉莉花茶辛温发散、芳香解郁，是立春茶饮的上选。

terms[1], which has both cultural considerations and health values.

除了依据季节选择茶品，我们也可结合个人喜好，在一天的不同时段选择不同的茶品。比如，清晨喝一杯淡淡的绿茶，醒脑清心；上午喝茉莉花茶，芬芳怡人，提高工作效率；午后喝杯红茶，解困提神；晚上与朋友或家人团聚，泡上一壶乌龙或普洱，别有一番情趣。

Based on personal preferences, you may also drink different teas at different times of a day. For example, you can drink a cup of light green tea in the morning to refresh a day; during morning break, a pleasant cup of jasmine tea helps improve work efficiency; black tea is effective when you start to get sleepy and tired in the afternoon; in the evening, it's fun to get together with friends or family over a pot of oolong or Pu'er tea.

日常茶饮

---

① Solar terms, also called Jieqi in Chinese, are days marking one of the 24 time buckets of the solar year in traditional Chinese calendar, and were used to indicate the alternation of seasons and climate changes in ancient China.

（2）饮茶禁忌 / Seven "Don'ts" in Drinking Tea

茶的保健作用毋庸置疑，但很多人缺乏对不良饮茶习惯的认知。日常生活中，喝茶有七个禁忌：空腹不喝茶，睡前不喝茶，用餐前后不宜多喝茶，不宜喝太浓的茶，不宜喝烫茶、冷茶，不喝冲泡时间过长的茶，不宜用保温杯泡茶。

While there is a major amount of positive results from drinking tea, there is one big problem—a lack of care for the side effects of drinking tea. There are seven "don'ts" in drinking tea, namely, don't drink tea on an empty stomach; don't drink too much tea before going to bed; don't drink much tea just before or after a meal; don't drink too much strong tea; don't drink the tea which is too hot or cold; don't drink tea brewed for a long time; and don't brew tea with a thermos bottle.

一忌空腹饮茶 / Don't Drink Tea on an Empty Stomach

数百年前，人们虽然不了解空腹饮茶时产生的物化反应，却已明确提出空腹喝茶伤肾、伤脾胃[①]。

Hundreds of years ago, although people did not understand the physicochemical reaction caused by drinking tea on an empty stomach, they clearly proposed that it was harmful to people's health.

现代研究表明，茶叶中的咖啡因对人体有一定的刺激和振奋作用。空腹饮茶会加速心率，导致心慌、头痛；且空腹饮茶会促进排尿，加重肾的负担。空腹喝茶太多，还容易出现"茶醉"现象，其症状表现

---

① 如清代黄宫绣在《本草求真》中就提出"至于空心饮茶，既直入肾削火，复于脾胃生寒，万不宜服。"

与晕车相似——头晕、乏力、心慌、想呕吐等。出现轻微的"茶醉"现象，不要紧张，只要口含糖果或者吃些点心，短时间便可缓解。

As we know, tea contains caffeine which stimulates central nervous system. Drinking tea on an empty stomach will accelerate the heart rate and even cause palpitations; besides, it will impose additional burden on the kidney. Too much tea on an empty stomach may result in "tea drunkenness" whose symptoms are similar to those of carsickness—giddiness, weakness, palpitation and nausea. In case of "tea drunkenness", eat some sweets or snacks and soon he/she will be recovered.

二忌睡前喝茶 / Don't Drink Too Much Tea Before Going to Bed

茶叶有利尿、提神、兴奋神经的作用，这主要是因为茶中有咖啡因成分。睡前摄入一定量的浓茶可能导致失眠、多梦等症状。此外，茶具有利尿功效也是睡前不喝茶的原因之一。

As we know, tea contains caffeine which has the functions of being diuresis and stimulating nerves. Having strong tea late at night may cause difficulties in falling asleep and a less restful sleep at night and it may force you get up at midnight to "answer the call of nature".

生活中有不少茶人，因长期饮茶，身体对茶的适应性较强，即使晚间喝茶也不会失眠。即便如此，为保障优质睡眠，也要适当控制睡前饮茶的量与浓度。

For some "heavy tea drinkers", drinking tea at night has no adverse effect. Even though, it is suggested that they drink less tea or lighter tea at night for the sake of their health.

三忌用餐前后大量喝茶 / Don't Drink Much Tea Just Before or After a Meal

饭前不宜多饮茶，否则会冲淡胃液，影响食物消化。饭后立刻大量饮茶，会影响铁和蛋白质的消化和吸收。

Do not drink tea just before meals. Drinking a lot of tea will dilute the gastric juice and interfere with digestion. Drinking tea right after a meal is also harmful, which may interfere with the digestion and absorption of iron and protein.

数百年前，古人就提出饭后立即饮茶伤脾胃[1]。当然，在这一点上，存在不同的文化习俗，如在欧美一些国家，饭后的那杯茶是用餐的完美句号[2]。

Ancient Chinese made it clear that drinking tea immediately after a meal was harmful to the spleen and stomach. However, in some European and American countries, people often drink a cup of tea after a meal.

四忌喝浓茶 / Don't Drink Too Much Strong Tea

浓茶的提神、兴奋神经的作用较强，故晚间饮用浓茶，容易影响睡眠。长期喝浓茶，牙齿有可能着色，严重的会发黄发黑。此外，浓茶影响身体对于钙、铁等元素的吸收，长期喝浓茶有可能导致骨质疏松和缺铁性贫血。总之，日常饮茶时我们应适当控制茶的浓度。

---

[1] 例如：《红楼梦》第3回"贾雨村夤缘复旧职，林黛玉抛父进京都"有这么一段："寂然饭毕，各有丫鬟用小茶盘捧上茶来。当日林如海教女以惜福养身，云饭后务待饭粒咽尽，过一时再吃茶，方不伤脾胃。今黛玉见了这里许多事情不合家中之式，不得不随的，少不得一一改过来，因而接了茶。早见人又捧过漱盂来，黛玉也照样漱了口。盥手毕，又捧上茶来，这方是吃的茶。"

[2] 很多欧美人都有饭后喝茶的习惯，这或许与他们的体质及其高糖、高热量、高蛋白的饮食有关。但饭后这杯茶，通常边喝边聊天，喝的速度慢且量也不多。

Many people consume strong tea to increase mental focus and improve energy. However, too much water drowned the miller. Drinking strong tea late at night may cause sleeplessness; it may also make one's teeth stained. Beside, strong tea may interfere with the absorption of calcium and iron, which may lead to osteoporosis and anemia. The solution is simple—do not make your tea too strong!

五忌喝烫茶、冷茶 / Don't Drink Tea Which is Too Hot or Cold

虽然适口的茶汤温度因人而异，从健康角度来说，茶汤温度控制在 50℃~55℃ 比较适宜。这是因为食管黏膜的耐受温度在 50℃~60℃，而刚沏好的茶水温度可达 80℃甚至 90℃。研究表明，茶水温度过高会损伤食道，长期饮用温度过高的茶会增加患食管癌和胃癌的风险[1]。

Although temperature is a matter of individual preference, warm tea proves to be a healthier choice. It is suggested that the temperature of tea be kept between 50℃–55℃. Many Chinese like drinking the freshly brewed tea which is very hot, reaching 80℃–90℃. Studies have shown that drinking hot tea is associated with a high risk of esophageal and gastric cancers.

茶太烫不好，太凉也不行。茶汤凉了之后，颜色会变浑浊、口感变差、香气受损，喝茶的怡然之情随之消散，更重要的是茶的保健效果大打折扣[2]。故饮茶最佳的温度不但取决于个人喜好，还应考虑身体的承受能力。

---

[1] 茶水温度过高，会烫伤食管黏膜，为了及时修补损伤，食道黏膜的上皮细胞就要加快增殖。如此反复就会引起黏膜质的变化，有可能进一步发展变成癌。研究表明饮用烫茶（70℃或更高）的人，患食管癌的风险是饮用温茶或凉茶的人的 8 倍。

[2] 需要指出的是，茶汤变冷与冷泡茶是不一样的。冷泡茶因温度始终较低，短时间内不会出现颜色加深、变涩变苦的现象。但冷泡茶常置于冰箱内保鲜，虽然爽口却并不适合所有人群。

For the sake of health, neither should people drink cold tea. If tea turns cold, the color will become dark, the taste will change and the aroma will be impaired. With a cup of cold tea, one gets little pleasure or health benefits. Therefore, though temperature is a matter of personal preference, we'd better drink tea at a proper temperature.

六忌喝冲泡时间过长的茶 / Don't Drink Tea Brewed for a Long Time

茶最好是现泡现喝，现泡的茶不仅滋味好，也更加卫生、有营养。冲泡时间过长，茶汤色暗、味差、香低，品尝价值大打折扣。以普洱茶为例，有些人冲泡普洱茶时，将茶沤在壶里，饮用时发现茶汤如酱油汤般难以入口；也有一些茶馆以普洱茶作迎客茶，在品饮时往少量"原汤"中注水稀释，虽口感尚可，"茶气"却寡淡了许多[1]。

Drink the tea when it is freshly brewed. Freshly brewed tea tastes good meanwhile it is healthy and nutritious. If the tea has been brewed for a long time, the infusion will lose its original color, taste and fragrance. Take Pu'er tea as an example, when the tea has been steeped in the pot for a long time, the infusion will get dark and bitter. In some teahouses, Pu'er tea is used as welcoming tea, that is to say, the tea is served free of charge before guests make their orders. When serving the tea, the waiter or waitress will pour a small amount of "original tea" into a pot and fill it up with boiling water. Although the taste is not that bad, the flavor is much poorer; let along the fragrance of the tea.

此外，茶也不宜过度冲泡。茶冲泡多次，茶叶中的维生素、氨基

---

[1] 这种冲泡方式与俄罗斯的茶炊、土耳其的子母壶的冲泡方式比较接近，口味尚可，韵味却少了许多。

酸等有益物质减少，茶汤营养价值大大降低。以绿茶为例，据测定，绿茶第一泡时，其可溶性物质能浸出 50%~55%，第二泡能浸出 30% 左右，第三泡能浸出 10% 左右，第四泡能浸出 1%~3%。这就是为什么人们常说绿茶三泡足矣。

In addition, tea cannot be brewed for too many times since the nutritional substances of tea is becoming less and less with each brew. Taking green tea as an example, for the first brew, its soluble substances can be leached by 50%–55%; for the second brew, about 30% of the soluble substances is leached; for the third brew, about 10% is leached; and for the fourth brew, only 1%–3% is leached. That is why green tea should be brewed for no more than three times.

从健康角度来说，传统的早起一壶茶，配上一壶水，从早喝到晚的方式需要改变了。

Therefore, the traditional practice of brewing a pot of tea and drink it for a whole day needs to be changed.

七忌用保温杯泡茶 / Don't Brew Tea with a Thermos Bottle

不少人在冬季会用保温杯泡茶，一则饮用方便，可以随时喝上热腾腾的茶水；二则不洒不漏，携带方便。保温杯泡茶虽然方便，却不太健康（尤其是绿茶）。一则保温杯保温效果太好，沸水泡茶，无法随泡随喝，尤其是杯盖无法当作茶杯使用时更是如此；二则口感香气变差，长时间高温浸泡会使茶多酚等浸出过多，茶汤色浓味苦；三则健康成分流失，长时间高温浸泡会破坏茶叶中的维生素和芳香物质，损害其营养成分。

Many people brew tea and keep it warm with thermos bottles in winter. In this way, they hope they can drink tea at any time and place.

Unfortunately, people soon find that thermos bottles let them down. First of all, the infusion in the bottles can be too hot to drink because of the good insulation, and if the lid cannot be used as a tea cup, you will regret for having such an excellent thermos bottle. Secondly, both the taste and aroma of tea change with time, due to the excessive leaching of tea polyphenols. Thirdly, longer soaking time leads to less nutrition, because vitamins and aromatic compounds are greatly damaged.

近几年出现的保温泡茶杯，虽说可解决茶水分离问题，控制茶汤浓淡，但冲泡好的茶汤经过一段时间的保温后，口感还是有所改变，其中绿茶茶汤的变化最为明显。

To solve the problems, thermos bottles for brewing tea have been invented in recent years. And it is highly praised for separating tea leaves from water and precisely controlling the brewing time. However, personal experience has proved that although the new invention is better than the traditional one, the effect is not as great as it is advertised, especially when green tea is concerned.

除了以上饮茶七忌外，还有些日常饮茶注意事项。如刚上市的春茶虽然金贵，但建议不要着急喝，因为刚刚炒制完成的茶叶未经放置，含有较多的未经氧化的多酚类等物质，容易导致肠胃不适；吃药前后两小时，或是在食用一些具有滋补效果的食物时，一般也不宜饮茶，以免影响药效。此外，万事"过犹不及"，切勿过量饮茶。

In addition to the above "don'ts", there are some other suggestions. For example, although the new spring tea is precious, it is recommended not to drink it right after it is produced, because it contains some substances which may cause gastrointestinal discomfort. Besides, do not drink tea two hours before or after taking medicine, or when you are having food with tonic properties. Though tea is safe and healthy, there are negative effects if one consumes too much[1].

当下，饮品市场不断推陈出新，新式饮品口味多元、外观时尚、方便快捷。在竞争激烈的新饮品时代，以奶茶为代表的新式茶饮独领风骚，这与茶的包容性有关，更与茶的健康功效紧密联系。相比于新式茶饮，纯茶通常更健康。

Today, the beverage market is constantly innovating, and new products attract more and more customers with diverse tastes, fashionable appearance as well as fast speed. In the fierce competition of beverages, tea-beverages represented by milk tea (Bubble tea or Boba tea) dominants the market. The dominance should be attributed to the inclusiveness of tea and the awareness of its health benefits. You should always keep in mind that the natural beverage is always a better choice than processed, sweetened beverages.

茶的种类丰富，功效亦有不同，我们应当充分了解自己，按需选茶、科学饮茶。让我们寻找适合自己的茶，开启健康饮茶之旅吧！

With thousands of different teas, you have endless choices. In the colorful world of tea, you may choose the ones you like and make good use of their properties. Let's start our tea journey!

---

[1] Drinking too much tea could lead to an iron deficiency because tea is rich in tannins, which can bind to iron and keep it from being absorbed. For tea drinkers, drinking too much tea can cause anxiety, restlessness or trouble sleeping.

小贴士：

1. 在提到《神农本草经》时，可向国外友人介绍：*Shennong Bencao Jing*（*The Divine Farmer's Herb-Root Classic*）was first compiled some time during the end of the Western Han Dynasty, about the time of Christ. 如此，便于外国友人了解该书的历史和重要性。

2. 在提到《本草纲目》时，可向国外友人介绍：With almost 2 million Chinese characters, *Bencao Gangmu,* also known as *Herbal Foundation Compendium*, lists 1,892 kinds of medical substances. Besides Chinese herbal medicines, it includes animals and minerals used as medicinal substances. It is so famous that it has been translated into more than 20 languages. Even today it is often used as a reference book.

3. 讲解中可以介绍近年来热门的草本茶或中医养生茶等。这些基本都是"非茶之茶"，因为它们包含草药、水果干、香料等多种成分，却大都没有茶叶。真正意义上的茶，必须来自茶树。It should be noted that herbal tea is not really tea because it contains many ingredients such as herbs, fruits and spices and the point is it contains no tea. To truly be tea, the leaves must come from the tea plant.

08 茶之馔
Paring Tea with Food

8.1 食饮相融 / Harmonizing Food and Drink

顾名思义，"餐饮"就是餐食与饮品的结合。在茶的故乡，作为饮品的茶与餐的关系可谓源远流长。早期人类将茶树的嫩叶直接咀嚼或煮食，作为补充营养和缓解疲劳的食物来源。随着茶叶加工技术的进步和饮食文化的发展，茶的食用方式逐渐演变为饮用，在这一演变过程中，唐代的《茶经》有着里程碑式的意义。在随后的一千多年里，作为饮品的茶与餐的关系并未因此终结，反而得到更多元的发展。

Catering, known as "Can Yin" in Chinese, means "food and drinks", highlighting the close relationship between the two. In China, the birthplace of tea, the connection between tea and food has a deep and enduring history. Early humans chewed fresh tea leaves or boiled them for sustenance and energy. As tea processing techniques advanced and culinary traditions evolved, tea gradually changed from being consumed as food to being enjoyed as a drink. A pivotal moment in this transformation was *the Classic on Tea*, written in the Tang Dynasty. In the centuries that followed, tea's relationship with food did not fade; instead, it thrived, taking on even more diverse and refined forms.

很多家庭在饭后围坐一堂，共享一壶香茗，既清新口腔又帮助消化，更添几分悠闲与惬意；亲朋好友相聚用餐，常以一杯热茶为开场白，为宴席拉开序幕，而在宴席即将落幕之际，"把茶言欢"则成为加深情感的温馨时刻。

In many Chinese households, families often enjoy a pot of tea after meals, which not only cleanses the palate and aids digestion but also fosters relaxation and a sense of togetherness. When friends and family gather for

把茶言欢（松 Chill 学院 供图）

a meal, tea often serves as the perfect prelude, creating a warm and inviting ambiance. As the meal winds down, tea again becomes a cherished ritual, enhancing the joyful atmospherc.

在广东、香港等地，茶餐厅遍地开花，"叹早茶"更是成为当地人生活中不可或缺的一部分。此外，蒙古族和维吾尔族的咸奶茶、瑶族和侗族的打油茶、藏族的酥油茶、土家族的擂茶，无不是茶与餐的绝妙融合。由此可见，茶餐的紧密联系，长久以来存在于中国人的日常生活中。

In Guangdong and Hong Kong, cha chaan tengs (tea restaurants) are ubiquitous, and "yum cha" (dim sum brunch) is an integral part of daily life[①]. Besides, different ethnic groups in China also have long incorporated tea into their culinary traditions. The Mongolians and Uyghurs enjoy salty milk tea; the Yao and Dong ethnic groups prepare oil tea; Tibetans savor yak butter tea; and the Tujia people make mashed tea—each a unique and flavorful fusion of tea and dining culture. All these traditions highlight the enduring bond between tea and food, a relationship that has remained deeply embedded in Chinese daily life for generations.

在西方，葡萄酒与餐食的搭配一直备受重视，许多高端西餐厅都设有侍酒师，根据菜肴和客人预算，为他们推荐合适的酒品。近年来，"茶配餐"的概念发展迅速，催生了侍茶师这一新兴职业，其主要职责

① "Cha Chaan Teng"是"茶餐厅"的粤语直译，指香港的茶餐厅。"茶餐厅"也可译为"Tea Restaurant"，泛指提供茶饮和简餐的餐厅。"Yum cha"是粤语"饮茶"的发音，在英语中通常指"going for dim sum"（去喝早茶，吃点心）。Yum cha 不仅仅是喝茶，还包括享用各类广式点心。相应的，"早茶"的翻译既可以用 Yum cha (dim sum brunch)，也可以用 breakfast tea (dim sum brunch) 等形式。因此处主要是描述粤语区茶饮，故采用粤语直译加注意译的形式。

是基于丰富的茶知识，结合客人选择的菜品和预算，推荐合适的茶品，并以贴心的服务提升整体餐饮体验。

In the West, the art of pairing wine with food has long been highly valued. Many high-end Western restaurants employ sommeliers who expertly recommend wines based on the dish selection and customers' budgets. In recent years, the concept of "tea pairing with food" has gained momentum, leading to the rise of the new profession of tea sommelier. Their primary role is to leverage deep tea knowledge to curate tea selections that complement the flavors of a meal while considering the customer's preferences and budget, ultimately enhancing the overall dining experience.

在中国，虽然茶与餐的紧密联系早已为人所识，但"茶配餐"作为一个新的消费概念还未形成气候。对比葡萄酒在西式餐饮中的地位，茶在中式餐饮中的存在方式显得相对低调。

In China, while the deep-rooted connection between tea and food has long been acknowledged, the concept of "tea pairing with food" as a structured consumer trend has yet to fully take shape. Unlike the prominent role wine plays in Western cuisine, tea in Chinese dining remains relatively understated.

8.2 茶配餐的地域性 / Regional Characteristics of Tea and Food Pairing

虽然"茶配餐"作为一个新的消费概念在中国算是萌芽初现，但如果我们从宏观角度审视，就会发现中国不仅是"茶配餐"的引领者，更是最佳实践者——中国地方饮食与茶的搭配是"茶配餐"的绝佳范例。

Although tea pairing is still in its infancy as a structured consumer concept in China, a broader perspective reveals that China is not only a leader in this practice but also its most exemplary practitioner. Arguably, the pairing of regional Chinese cuisine with tea serves as a perfect embodiment of tea pairing in its most authentic and refined form.

一方水土一味茶，餐食习惯决定饮茶口感。戎新宇在《茶的国度：改变世界进程的中国茶》中提出，中国的江南 [①] 鱼米之乡，总体饮食口味趋于清淡，最适宜饮用清淡的绿茶，因而西湖龙井、洞庭碧螺春、安吉白茶这类鲜爽度高的绿茶颇受江南人的偏爱。

Generally, tea preferences are largely shaped by dietary habits. The tea expert, Rong Xinyu, suggests in the book *Tea Nation* that the overall diet in Jiangnan—a region renowned for its abundance of fish and rice—tends to be light and delicate, making it particularly suited for mild green teas. As a result, fresh and subtly flavored teas such as Xihu Longjing tea, Dongting Biluochun tea, and Anji White tea are highly favored by the people of Jiangnan.

武夷山虽地处福建，但当地人祖上多为江西移民，饮食习惯深受江西菜的咸辣风味影响，这种以咸辣为主的重口味与高发酵、重焙火的武夷岩茶相得益彰。

Meanwhile, in northern Fujian, the famous Wuyi Yan Cha (Wuyi Rock Teas) reflects the region's culinary influences. Despite being geographically located in Fujian province, many residents of Wuyi Mountain trace their roots to Jiangxi, a province known for its bold, salty, and spicy cuisine. This flavor profile naturally complements the roasted notes of highly fermented and baked Wuyi Yan Cha.

---

① 此处的江南为狭义上的江南，指上海、苏南、浙北、皖南等区域。

广东潮汕地区饮食口味并不重，本应饮绿茶较为适宜。但该地食材以海鲜居多，而海鲜在中国人的传统饮食观念中偏于寒凉。绿茶含有较多的茶多酚类物质，对肠胃消化刺激性较强，也被视为寒凉之物。为了避免寒上加寒的双重刺激，潮汕地区创制出适合当地饮食结构的轻发酵、重焙火的凤凰单丛茶。

The cuisine of the Chaoshan region in Guangdong province is relatively mild in flavor, which would typically make green tea a suitable choice. However, seafood is a staple in this area, and in traditional Chinese dietary principles, seafood is considered a "cooling" food. Green tea, rich in tea polyphenols, also has a strong stimulating effect on digestion and is similarly classified as a cooling substance. To avoid the excessive cold-on-cold effect, the Chaoshan region developed lightly fermented yet heavily roasted Fenghuang Dancong tea (Phoenix Single Bush tea), which better complements the local diet.

西藏、新疆、青海、内蒙古等以游牧民族居多的地区，其日常饮食以肉类及奶制品为主，瓜果蔬菜相对匮乏，这样的饮食结构就适合饮用纤维素含量较高，可促进代谢、消食解腻的黑茶 [①]。

In regions such as Xizang, Xinjiang, Qinghai, and Inner Mongolia, where nomadic communities are predominant, daily diets are heavily based on meat and dairy products, while fruits and vegetables are relatively scarce. This dietary structure pairs well with black tea, which has a high cellulose content and helps promote metabolism and digestion.

---

① 关于餐食习惯决定饮茶口感部分的论述，源自戎新宇的《茶的国度：改变世界进程的中国茶》，在表述上略有简化和修改。

8.3 茶配餐的常见形式 / Common Forms of Tea and Food Pairing

（1）以茶佐膳 / Pairing Tea with Food

如果说餐食习惯决定饮茶口感是茶配餐的宏观表现形式，那么在生活中茶与餐的搭配就是其具体表现形式。换句话说，在用餐过程中，根据菜品选择茶品是茶配餐最直接的表现。[①]

If dietary habits shape tea preferences as a macro-level reflection of tea pairing, then the everyday combination of tea and food represents its practical application. In other words, selecting the right tea to accompany a meal is the most direct expression of tea pairing.

正如葡萄酒与餐食的完美搭配能让二者相辅相成，提升用餐体验，茶与餐食的巧妙搭配也能相互增色，提升整体的味觉体验。若搭配得当，茶能衬托出食物的美味；若搭配不当，菜肴可能会掩盖茶的韵味，或者茶影响菜肴的口感。

Just as a well-matched wine and dish can complement and elevate the dining experience, a thoughtfully paired tea can also enhance flavors and create a more refined dining experience. When paired correctly, tea can bring out the best in a dish; if mismatched, the food may overpower the tea, or the tea may interfere with the dish's intended flavor.

---

① 有人认为边吃饭边喝茶不利于健康，因为茶叶含有草酸，会影响营养吸收，茶水会冲淡胃酸，影响消化。其实菠菜、芹菜等蔬菜中草酸的含量是茶的很多倍，只要注意正确的饮用方法，控制好饮茶量和茶水浓度，无须太过担心。

　　从茶饮与菜肴的风味搭配规律来看，大致有两种搭配模式。当茶与食物滋味迥异或差异显著时，两者相配可平衡味觉冲突，创造新的味觉维度。例如，用清苦的绿茶搭配高甜度的点心，绿茶中的茶多酚具有收敛性，可中和甜腻，形成清爽回甘的余韵。当茶饮与食物呈现相似风味基调时，两者相配可增强味觉记忆。例如，普洱熟茶的陈年木质香与烤羊排的焦香形成复合香气，既能解腻增鲜，又能凸显肉质的鲜美。

　　From a flavor-pairing perspective, tea and food can generally be combined in two ways. The first is contrast pairing, which means when tea and food have distinct or contrasting flavors, they can balance out taste differences and introduce new layers of complexity. For example, pairing bitter green tea with sweet dim sum—the astringency of tea polyphenols helps counterbalance the sweetness, leaving a refreshing, lingering aftertaste. The other is complementary pairing, which means when tea and food share similar flavor notes, good pairing can reinforce and intensify the taste memory. For instance, the aged, woody aroma of ripened Pu'er tea and the charred

以茶佐膳（松 Chill 学院　供图）

smokiness of grilled lamb chops combine to form a complex aroma that not only cuts through the richness but also enhances the umami of the meat.

当然我们也要避免过犹不及，即避免使用风味完全相同的茶饮与菜肴组合。例如，用桂花乌龙茶配甜腻的桂花糯米藕，会导致味觉疲劳。

However, overmatching should be avoided—pairing tea and food with identical flavors may lead to palate fatigue. For example, drinking osmanthus oolong tea alongside osmanthus-flavored sticky rice cake can result in an overwhelming, monotonous taste experience.

从茶类特征角度来看，带有大自然草本植物香的绿茶，适合搭配一些味道清爽的蔬菜、沙拉、凉菜，也可搭配较为清淡的鸡肉。乌龙茶的滋味丰富、香气浓郁，通常会有宜人的焙火味，既能衬托出烟熏肉类的味道，也适合搭配一些口感厚重的菜肴。红茶浓郁甘甜，适合搭配"硬菜"，如烤牛肉、烤羊肉和烤鹿肉等。白茶的毫香宜人、清淡回甘，可与同样清爽的鱼类、乳制品、蔬菜和甜品等搭配。黑茶（如熟普）醇厚，与炖煮类荤菜搭配，去腻增香；与重口味卤味搭配，解咸提鲜。

From the perspective of tea characteristics, different teas pair well with different types of food. Green tea, with its natural herbal fragrance, is best suited for light, refreshing dishes such as vegetables, salads, cold appetizers, or delicate chicken dishes. Oolong tea, known for its rich taste and roasted aroma, complements smoked meats and dishes with dense textures. Black tea, with its full-bodied and sweet profile, pairs well with hearty meats like grilled beef, lamb, and venison. White tea, with its delicate fragrance and mild aftertaste, is ideal for light seafood, dairy products, vegetables, and desserts. Dark tea (such as ripened Pu'er tea), known for its smooth, mellow taste, can be paired with stewed meats to cut through the grease and enhance aroma, or with rich, braised dishes to balance saltiness and boost freshness.

茶餐搭配得当不仅能提升用餐体验，还能让茶与餐相互成全，带来美妙的味觉体验。

A well-executed tea pairing not only enhances the dining experience but also deepens the appreciation of both the tea and the food, creating a truly harmonious blend of flavors.

（2）以茶入膳 / Incorporating Tea into Food

除了以茶佐膳外，在茶为国饮的中国，人们很早就开始用茶来料理美食。从某种意义上看，这是茶配餐更深入的表现形式。

In China, where tea is regarded as the national drink, tea is not only enjoyed as a drink but has also been incorporated into cuisine for centuries. In many ways, this represents a deeper and more integrated form of tea pairing.

以茶入膳，主要取茶叶的香气和滋味，使其与食物完美结合。鉴于茶叶具有一定的药理成分，茶叶菜肴（或称茶膳）一般都具有双重功效，既可增进食欲，又有一定的保健效果。

Tea-infused cuisine, also known as tea-based meals, highlights the natural aroma and flavor of tea, seamlessly integrating it into various dishes. Rich in bioactive compounds, tea-infused cuisine is often valued for its ability to enhance appetite while offering potential health benefits.

以茶入膳可分为茶菜肴、茶主食、茶汤、茶点心等。茶主食就是将茶以各种形式添加到主食中，常见的有茶米饭、茶面条、茶饺、绿

抹茶月饼

茶粥、茶叶蛋等。茶汤是指在羹汤中佐以各式茶或茶香，常见的茶汤有绿茶蛋汤、红茶银耳汤等。茶点心是将茶添加到点心中，常见的茶点心有抹茶蛋糕、抹茶冰激凌、红茶饼干等。

Tea-infused cuisine can be categorized into tea staples, tea soups, tea snacks and tea-based dishes. Tea staples incorporate tea into main dishes in various ways, such as tea-flavored rice, tea noodles, tea dumplings, green tea congee, and tea-marinated eggs. Tea soups are soups infused with tea leaves or tea essence. Common examples include green tea egg soup and black tea snow fungus soup. Tea snacks include tea-flavored cakes, tea-flavored ice cream, and black tea biscuits.

茶味冰激凌

红茶凉粉

作为以茶入膳最重要的形式，茶菜肴是将茶叶作为辅料入菜，一般有四种方式，包括鲜叶入菜、茶汤入菜、茶粉入菜和茶香熏菜。

Tea-based dishes is the most significant form of tea-infused cuisine, where tea leaves are used as an ingredient to complement the flavors of a dish. There are four common methods: incorporating fresh tea leaves, using tea broth, adding tea powder, and tea-smoking the food.

茶菜肴中最知名的非杭州名菜龙井虾仁莫属，此道菜以清明前后的龙井配以虾仁制作而成，龙井茶既解了虾仁的腥味，又留下满口茶香；绿茶香气清新，入口清爽，最适合烹煮此类淡雅的菜肴^①。乌龙茶滋味香浓醇厚，具有健胃消食之功效，适合与重油的菜肴搭配，如大红袍炖排骨，茶香浓郁、酥烂入味。红茶香甜味醇、去腥解腻，适合与口味重、色泽深的菜肴搭配；用红茶炖牛肉，去腥增鲜、滋味醇厚。普洱老茶香气持久、滋味浓醇，适合做卤水汁，用于焖、烧效果最好，如普洱炖排骨、普洱茶卤鸡等。

龙井虾仁

茶香排骨

One of the most famous tea-based dishes is Fried Shrimps with Longjing Tea, a delicacy from Hangzhou. This dish is prepared with freshwater shrimp and the renowned Longjing tea harvested around the Qingming Festival. The tea not only removes the shrimp's fishy odor but also imparts a delicate tea

① 如绿茶拌豆腐，将豆腐焯水，加油、盐等配料，再加提前泡开的绿茶（如龙井茶）即可。此菜清凉爽口，适合春夏食用。

fragrance, enhancing its fresh and light flavor. Green tea, with its fresh aroma and taste, is ideal for such delicate dishes and light stir-fries. Oolong tea, known for its rich and mellow profile, aids digestion and is well-suited for heavier, oil-rich dishes. Da Hong Pao Braised Pork Ribs are infused with a rich tea aroma, delivering tender, fall-off-the-bone meat with deep, savory flavors. Black tea, with its naturally sweet and smooth taste, effectively neutralizes fishy and greasy odors, making it an excellent complement to stewed dishes like black tea braised beef. Pu'er tea, known for its mellow and aged aroma, is well-suited for braising and stewing, often featured in dishes such as Pu'er tea braised pork ribs or Pu'er tea braised chicken.

以茶入膳，贵在保留茶香，因此精准掌握火候是制作茶膳的关键。此外，葱、姜、蒜这些重口味的佐料会影响茶香，做茶膳时尽量少放。从茶类来看，绿茶入菜以清炒较好，乌龙茶、红茶、普洱茶等发酵类茶入菜以炖煮为宜。

The key to making tea-based dishes lies in preserving the tea's natural aroma, making precise heat control crucial. Additionally, strong seasonings like onions, ginger, and garlic can easily overpower the tea's fragrance, so they should not be used in tea-based recipes. From a culinary standpoint, non-fermented green tea pairs best with stir-fried dishes, while fermented teas such as oolong, black tea, and Pu'er excel in braising and stewing, enhancing depth and richness in flavor.

（3）以膳佐茶 / Pairing Snacks with Tea

一般而言，在茶配餐中，餐是主体，茶是辅助；而在茶馆、茶空间等场所，茶与餐就要发生角色互换，茶自然地成为主角。这里的餐多指茶点，即佐茶的点心。茶点通常精细美观、形小量少、口味多样、品种丰富。

In tea pairing, food generally takes center stage, with tea serving as a complementary element. However, in tea houses or tea clubs, this dynamic is reversed—tea becomes the star, while food plays a supporting role. In these settings, food primarily refers to tea accompaniments, commonly known as tea snacks. These delicate, artfully crafted treats are typically served in small portions, offering a diverse array of flavors and a variety of choices.

常见茶点可分为四类：干果类、鲜果类、糖果类和点心类。干果类通常包括瓜子、花生、开心果、山楂条、杏干、葡萄干等；鲜果类包括橙、苹果、香蕉、提子、菠萝、猕猴桃等；糖果类包括芝麻糖、花生糖、牛轧糖等；点心类包括蛋糕、曲奇饼等西式点心，烧卖、虾饺、豌豆黄、绿豆糕、凤梨酥等中式点心，还包括近年来比较流行的中西合璧式点心（中点西做）[1]。

Tea snacks can generally be categorized into four main types: Dried fruits and nuts, such as melon seeds, peanuts, pistachios, hawthorn strips, dried apricots, and raisins; Fresh fruits, such as oranges, apples, bananas, grapes, pineapples, and kiwis; Candies, such as sesame candy, peanut candy, and beef jerky candy; Dim sum, a broad category that includes both Western-style pastries (e.g., cakes, cookies) and traditional Chinese dim sum (e.g., shaomai or shumai, shrimp dumplings, pea puree cakes, bean paste cakes, pineapple cakes). In recent years, a fusion of Chinese and Western dim sum has also gained popularity.

---

[1] 如熔岩绿豆糕，即以流心巧克力工艺重构绿豆馅，高温烘烤后形成"爆浆"效果，兼具传统豆香与西式甜点的美味；再如桂花慕斯糕，即以桂花蜜融入西式慕斯层，搭配绿茶冻夹心，打造晶莹剔透的立体造型，在保留中式清香的同时，增强口感的绵密性。

茶点种类繁多、滋味丰富，选择时不仅需要考虑茶点与茶品相配，还应考虑品饮的季节、地区饮食特点，同时兼顾消费者个人的喜好。就茶品而言，一般遵循"甜配绿、酸配红、瓜子（干果）配乌龙"。

With such a wide selection of tea snacks, choosing the right pairing requires more than just matching flavors—it also involves considering seasonality, regional dietary preferences, and personal taste. A general rule of thumb is "sweet snacks pair well with green tea; sour snacks complement black tea; and sunflower seeds go best with oolong tea."

就季节而言，春季可选色彩鲜艳的，夏季准备相对清淡的，秋季可选时令鲜果，冬季的茶点口味可以偏重。

广式茶点

When considering seasonal pairings, different flavors and textures complement different seasons of year. In spring, bright-colored snacks reflect the season's vibrancy and freshness. In summer, lighter and more refreshing options help counter the heat. In autumn, fresh fruits make for a natural and harmonious pairing. In winter, richer and bolder flavors provide warmth and depth, creating a more comforting experience.

以餐佐茶（松 Chill 学院　供图）

就地区特色而言，可选择地方特色点心，如在北京可选豌豆黄，在江南可选绿豆糕，在台湾可选凤梨酥。

To enhance authenticity, we may select local specialties that complement the tea experience. Examples include pea puree cake from Beijing, bean paste cake from Jiangnan, and pineapple cake from Taiwan.

佐茶点心

除考虑以上因素外，也可以结合个人喜好进行选择。

While these guidelines provide a helpful framework, individual taste preferences should always take precedence in creating the perfect pairing.

8.4 茶配餐的创新模式 / Innovative Models of Tea Pairing

新式茶饮的主要原料除了茶叶之外，一般会有鲜奶、水果、芝士、木薯、芋艿、坚果碎甚至饼干等丰富食材。更为重要的是，为拓宽发展路径，提高客单价，几乎所有热门茶饮品牌都开发了烘焙业务线，利用茶饮解腻特性搭配烘焙品的甜香，故新式茶饮可被视为茶配餐的创新模式。

Incorporating a diverse range of ingredients such as fresh milk, fruits, cheese, tapioca, taro, crushed nuts, and even cookies, new-style tea drinks have evolved far beyond traditional tea. More importantly, to expand their market reach and increase customer value, nearly all major tea beverage brands have ventured into the baking industry, leveraging tea's natural astringency to complement the rich sweetness of baked goods. As a result, new-style tea drinks have emerged as an innovative model of tea pairing.

融合了传统茶文化与现代创新元素的新式茶饮，以其独特的口感、丰富的种类和时尚的包装，丰富了人们的饮品选择，赢得了消费者的喜爱。与传统茶饮相比，新式茶饮在注重茶底的选择的基础上，尤其强调搭配和创意。据统计，当前头部品牌通常单杯使用3种以上配料。常见配料包括珍珠、芋圆、麻薯、芋泥等淀粉类，芝士、布丁等乳脂类，血糯米、燕麦粒等谷物类，烧仙草、龟苓膏等草本类。在新式茶饮的世界里，只要是美味的、安全的，一切皆可入茶。

创意茶饮

By blending traditional tea culture with modern creative elements, these new-style tea drinks stand out for their unique flavors, wide variety, and trendy packaging. Unlike traditional tea drinks, which focus primarily on the tea base, new-style tea drinks emphasize ingredient combinations and creative pairings. Statistics show that today's top brands incorporate an average of three or more ingredients per cup, with common additions including starches (e.g., tapioca pearls, taro balls, sweet potato cubes), cream-based toppings (e.g., cheese foam, pudding), grains (e.g., blood glutinous rice, oatmeal), herbal ingredients (e.g., grass jelly, a herbal jelly called Guilinggao). In a word, in the world of new-style tea drinks, if an ingredient is delicious and safe, it can be infused into tea.

新式茶饮不仅满足了人们对健康、美味、便捷、时尚饮品的需求，更成为一种生活方式的象征。从繁华的都市街头到宁静的小镇巷尾，新式茶饮店如雨后春笋般涌现，成为人们休闲、聚会、打卡的热门去处。

While offering healthy, flavorful, and convenient beverages, new-style tea drinks have transformed the tea-drinking experience. From bustling city streets to quaint town alleys, new-style tea shops have sprung up like mushrooms after the rain, becoming popular social hubs for relaxation, gatherings, and social media check-ins.

此外，新式茶饮的品牌营销实现了线上线下的全面融合。通过社交媒体、电商平台等渠道，新式茶饮品牌能够迅速传播品牌形象，吸引消费者关注。这种全球化的传播方式，也让新式茶饮走出了国门，成为中国茶文化的一张亮丽名片。

Moreover, new-style tea brands have seamlessly integrated online and offline marketing strategies. By leveraging social media, e-commerce platforms, and digital advertising, they have successfully built brand awareness and engaged consumers. This globalized marketing approach has propelled new-style tea drinks beyond China, establishing them as a symbol of modern Chinese tea culture on the world stage.

8.5 茶配餐的发展趋势 / Trends in Tea Pairing

随着消费者健康意识的提升，未来茶与餐的结合，无论是以茶佐膳、以茶入膳，还是以膳佐茶，都会越来越注重健康。同时，二者的结合不仅局限于产品本身，还将融入消费者的生活场景，强调品质化、个性化的消费体验。

192

As consumer awareness of health and wellness continues to grow, the future of tea pairing—regardless of its form—will increasingly prioritize health-conscious choices. Beyond just product innovation, the integration of tea and food will seamlessly blend into everyday lifestyle scenarios, focusing on quality, personalization, and immersive dining experiences.

在新式茶饮蓬勃发展的当下，很多餐厅抓住契机，推出或创制品牌专属的新式茶饮，如主打"火锅＋茶憩"的品牌凑凑早在 2016 年就打造了专属茶饮品牌"茶米茶"，并成为品牌面对竞品的差异化撒手锏；2018 年，火锅品牌小龙坎开设名为"龙小茶"的茶饮店；2021 年，海底捞推出首家 9.9 元自助奶茶。

With the rapid rise of new-style tea drinks, many restaurants have seized the opportunity to launch their own exclusive in-house tea brands. In 2016, the popular hot pot brand COUCOU introduced its tea brand "Teametea", which has become a key differentiator in an increasingly competitive market. In 2018, another hot pot brand, Shoo Loong Kan, launched "The Loong Tea", further expanding its offerings. In 2021, Haidilao, the leading hot pot chain, rolled out its first self-service milk tea, priced at just 9.9 RMB, blurring the lines between tea and dining.

此外，茶配餐领域还出现品牌强强联手的联名产品[1]，如 2022 年，湘菜品牌费大厨联手新式茶饮头部企业喜茶推出"超解辣杨梅冻"[2]。

---

[1] 茶配餐领域联名产品包括奈雪的茶与知名冰激凌品牌哈根达斯联名推出的"冰激凌茶饮系列"、奈雪的茶与海底捞推出的联名饮品"霸气雪顶山楂草莓"和"霸气楂楂和牛慕斯"等。

[2] "超解辣杨梅冻"的茶底选用茉莉花茶，配上杨梅鲜果和弹弹冻，与辣椒炒肉搭配，以茶品"解辣"菜品，彼此配合又彼此成就。

Beyond in-house tea innovations, strategic brand collaborations have also gained momentum in the tea pairing industry, such as the 2022 partnership between leading tea brand HEYTEA and Hunan cuisine brand Chef Fei, which launched the "Super Spicy Relief Waxberry Jelly".

在餐厅积极发展茶饮的同时，不少茶馆或新式茶饮品牌也在探索"茶+餐"的新模式。"茶+餐"可以拓展消费场景，延长经营时段，提高整体收益。如知名茶饮品牌奈雪的茶将新式饮品和软欧包结合，达到"1+1>2"的效果；主打"茶饮+餐饮"的因味茶（inWE），店内配有丰富的中西简餐和甜点。

While restaurants are actively expanding their tea beverage offerings, tea houses and new-style tea brands are also exploring the "tea + dining" model as a new business strategy. This approach expands consumption scenarios, extends business hours, and drives higher revenue. For instance, Naixue, a premium tea brand, has created a "1+1>2" synergy by pairing tea beverages with soft European-style bread. Meanwhile, InWE Tea seamlessly integrates tea and dining, offering a curated selection of Chinese and Western light meals and desserts, transforming tea drinking into a complete dining experience.

很多城市都有以售卖"茶+自助餐"为特色的茶馆，在这种模式下，茶与餐的搭配完全取决于客户的喜好。近年来，除了一些高端茶宴外[1]，一些中端茶馆陆续推出茶餐套餐。在此种模式下，店家的专业度体现在茶与餐的绝妙搭配上，比如在回锅肉套餐中搭配铁观音、在清蒸鲈鱼套餐中搭配绿茶、在红烧肉套餐中搭配红茶等；这种模式最

---

[1] 如创办于1994年的上海秋萍茶宴馆是中国首家茶宴馆，以"西湖十景"和"经典古诗宴"等特色茶宴闻名；位于上海外滩源的米其林一星餐馆——逸道茶餐馆，其菜单融入茶饮搭配建议，部分菜品更是以茶汤调味。

大的优势就是方便省事，吃、喝、谈事聊天一条龙，无须转场。

Today, many cities are witnessing the rise of "tea + buffet" tea houses, where customers can personally curate their tea and food pairings, creating a highly customized and immersive experience. In addition to luxury tea banquets, mid-range tea houses are increasingly offering tea meal sets, showcasing expertly crafted tea-food pairings. Examples include Tieguanyin Tea with Twice-Cooked Pork, Green Tea with Steamed Perch, and Black Tea with Braised Pork—each enhancing the flavors of the other. The biggest advantage of this model lies in its convenience and accessibility, offering a one-stop dining experience where guests can eat, drink, and socialize in a single venue without the need to move between locations.

茶餐套餐（松 Chill 学院　供图）

当然，"茶 + 餐"的消费场景还可以进一步拓展，比如有些茶空间除了会根据季节、客户偏好及所点菜品推荐茶品外，还将茶与素食、瑜伽相结合，串联多个消费场景。此外，将茶、餐、会务、茶服（购物）、茶疗（休闲）、古琴（培训）等元素结合起来的一站式休闲茶空间也应运而生。

As the concept of "tea pairing with food" continues to expand, some tea houses now offer personalized tea recommendations, carefully tailored to seasonality, customer preferences, and dish selection. Others integrate tea into wellness-focused experiences, such as vegetarian cuisine, yoga retreats, and mindfulness sessions. Additionally, multi-functional tea spaces are emerging, combining tea, fine dining, conference services, tea fashion (shopping), tea therapy (relaxation), and guqin (classical music training)—creating a truly holistic leisure experience.

习茶空间

随着人们对健康生活方式的追求、对中国传统文化的喜爱、对创新与时尚的追求，相信蕴含无限创意与无尽美妙的茶餐搭配必将吸引更多人的目光与味蕾。

As more and more people embrace healthy lifestyles, traditional Chinese culture, and innovative dining trends, the boundless creativity and exquisite flavors of tea and food pairings are sure to captivate more hearts and palates.

第四部分
茶香四海
Connecting the World with Tea

09 茶之传
Globalization of Tea

　　中国是茶的故乡，茶随着丝绸之路去了不同的国家，200 多年前茶又开启了全球化发展历程。今天，茶是世界三大无酒精饮品[1]之一。全球产茶国和地区达 64 个[2]，有喝茶习惯的国家和地区有 160 多个，饮茶人口超过 30 亿。

　　Tea used to be a unique natural resource in China. It was first brought to different places of the world along the Silk Road and then spread across the world over 200 years ago. Today, tea is one of the three major non-alcoholic drinks in the world. It is produced in 64 countries and regions worldwide,

---

[1]　世界三大无酒精饮料包括茶、咖啡和可可。

[2]　世界茶叶生产主要分布在亚洲和非洲，据国际茶叶委员会统计，其中，亚洲占 80%，非洲约占 13%，其他地区只约占总产量的 7%。亚洲茶叶产地主要集中在中国、印度、斯里兰卡和印度尼西亚等国，上述四国茶叶产量占亚洲茶叶总产量的 83%；非洲茶叶产地分布在肯尼亚、马拉维、乌干达等国。

茶山景致（图虫　供图）

and enjoyed by over 3 billion people, from over 160 countries and regions.

目前，在世界范围内对茶的称呼，大致可分为"Cha"和"Te"两大类。"Cha"系列的发音来源于中国北方官话，影响到南亚、西亚、中亚及欧洲小部分地区，最远传到非洲东部；"Te"系列的发音来源于中国闽南等地方言，伴随着茶叶的大量出口，闽南方言中茶字的发音在东南亚、非洲南部和西欧传播开来。茶的不同发音为我们清晰地绘制出茶叶在历史上的传播路径，并勾勒出全球极为重要的两条贸易路线："Cha"系列的是陆上丝绸之路，"Te"系列的则是海上丝绸之路。

Though every country has its own word for tea, almost all pronunciations stem from two root words: "Cha" and "Te". The pronunciation of the "Cha" series originated from standard Chinese in northern China, and it has been adopted by many countries in South Asia, West Asia, Central Asia, and some parts of Europe; besides, it has even influenced countries in the east of Africa. The pronunciation of the "Te" series originated from dialects in coastal areas of southern Fujian province. As tea was exported along maritime Silk Road, so was the pronunciation. That was why in Southeast Asia, southern Africa, and Western Europe, tea was pronounced like "Te". Just from the pronunciation, we can roughly know how tea was introduced to a certain country. In other words, if tea was introduced to a country along the Silk Road, it would be pronounced as "Cha"; if it was introduced along the maritime Silk Road, it would be referred to as "Te".

9.1 日本 / Japan

饮茶最初是一种使僧人在长时间打坐时保持清醒的方法。茶早期在日本的传播与发展始终与佛教息息相关。

The connection of tea and Buddhism is proverbial and it is long believed that in the early days tea was drunk by monks to keep awake while practicing meditation. So it may not be surprising that tea was introduced to Japan mainly by monks.

近水楼台先得月，日本早在8世纪（甚至更早）就知道了茶[1]，最早传入日本的茶很可能是遣唐使带去的。805年，日僧最澄[2]从中国带回茶籽，种在日本日吉大社旁边，是茶种传入日本的最早记载。815年，留学中国的僧人空海回到日本后不久，便留下了日本最早的饮茶记录[3]。

---

① 也有人提出，早在6世纪随着佛教从中国传入日本，日本人就知道了茶的存在。
② 最澄（767—822），日本天台宗的开创者，曾赴中国学习佛法。
③ 空海给天皇上了一份《空海奉献表》，其中说道："……茶汤坐来，乍阅震旦之书"，这是日本最早的饮茶记录。

1191 年，赴中国研习禅法的荣西禅师①，将茶种及宋代种茶、制茶和饮茶的知识带回了日本。荣西禅师撰写了日本第一部与茶有关的专业书籍《吃茶养生记》，不久抹茶为寺庙和上层社会所普遍享用。

Following in the footsteps of neighboring China, Japan got to know the Chinese tea and its culture in the 8th century②. During the Tang Dynasty, probably it was ambassadors to the Tang Court who brought tea and its knowledge back to Japan. It is believed that in 805, a Japanese monk named Saicho brought some tea seeds back to Japan and planted them near a temple, which was the oldest tea garden in Japan. The earliest record of tea drinking in Japan was written by another monk named Kükai visiting China in 815. In 1191, tea seeds, tea planting and processing techniques③ and associated social practices of the Southern Song Dynasty were brought back to Japan by Monk Yeisai-zenji, who studied Zen Buddhism in China. He also wrote the first tea book in Japan—*Drinking tea for Nourishing Life*. Soon the custom of drinking matcha (similar to the whipped tea) got popular in temples and upper classes.

15—16 世纪，经过村田珠光、武野绍鸥、千利休三位大师的努力，饮茶更为普及，成为集日常生活与哲学于一体的茶道。17 世纪，叶茶泡饮法传至日本，被称为煎茶。煎茶④很快就取代抹茶成为最受欢

---

① 明庵荣西（1141—1215），在中国学习佛法多年，他将带回日本的茶种成功种在了三处茶园，其中一处是京都附近的宇治，多年后宇治以世界顶级产茶地而闻名。

② Some experts claim the knowledge of tea was brought to Japan in the 6th century at the same time as Buddhism.

③ The seeds were planted in three places, and later one of them, the Uji district near Kioto, got famous for producing the best tea in the world.

④ 17世纪，中国高僧隐元禅师东渡弘法，将"瀹茶法"传到日本。当时京都宇治的制茶师将抹茶的蒸青制法与中国炒青的揉捻工艺结合，制作出蒸青煎茶，成了日本煎茶道的滥觞。

迎的日常茶饮，但这丝毫没有动摇抹茶在日本茶文化中的地位。

In the 15th and 16th centuries, Murata Shuko, Takeno Joo and Sen Rikyu made great contributions to the popularization of tea and its culture, and the Japanese Tea Ceremony was established as an independent art form. In the 17th century, loose leaf tea or steeped tea was brought to Japan. Known as Sencha, it soon replaced the powdered matcha tea in ordinary consumption, though the latter still remains its place as the tea of teas.

日本茶道（视觉中国　供图）

纵观日本饮茶的千年历史可以发现，其饮茶方式的变化是随着中日茶文化交流史上的三次高潮而来。8 世纪，遣唐使带回了煮饮的方式；12~13 世纪，以荣西为代表的僧侣们带回了冲饮的方式，后形成了抹茶道文化①；17世纪，中国的叶茶泡饮法传至日本，后形成了煎茶道文化。茶及其文化在日本本土化过程中，创造了契合日本民族特性且极富日本特色的表现形式——茶道，成为日本文化最具代表性的一面。

From a historical point of view, Japan closely followed the footsteps of Chinese tea in its three stages. Boiled tea was brought to Japan in the 8th century by ambassadors to the Tang Court; powdered tea was brought back in the 12th and 13th century by monks; loose leaf tea and the way of steeping tea for drinking was introduced to Japan in the 17th century. Finally, the ritual of preparing and serving Japanese powdered green tea was well established. Called Teaism, Tea Ceremony, Chanoyu, Sado or simply Ocha in Japanese, the Japanese Tea Ceremony has been the best showcase of Japanese culture②.

9.2 朝鲜半岛 / Korean Peninsula

朝鲜半岛饮茶有近1500年历史。与日本类似，茶最先传入朝鲜半岛也与佛教有关。茶于7世纪传入朝鲜半岛③，彼时茶被局限于祭祀和礼佛等重要场合。至新罗统一时期（668—935）以及高丽时期（935—1392），茶的逐渐普及依然与佛教的传播紧密关联。新罗时期是茶文化的第一个发展时期，开始了朝鲜半岛本土茶叶种植的历史。

---

① 现在的日本茶道分为抹茶道与煎茶道两种，但"茶道"一词所指的是较早发展出来的抹茶道。

② 为方便理解，日本茶道的英文统一采用"the Japanese Tea Ceremony"表述。

③ 《三国史记·新罗本纪》有记载："……入唐回使大廉持茶种子来，王使命植于地理山。茶自善德王有之，至于此盛焉。"由此可知，朝鲜从新罗善德女王时期（632—647）开始饮茶；遣唐使金大廉于828年从唐朝带回茶种。

彼时的饮茶方式仿效唐代煎茶法，饮茶基本集中于上层社会、佛教寺庙，也逐渐向民间普及。高丽时期是茶文化的兴盛期①，饮茶方式上，初期以煎茶为主，中后期流行点茶；高丽王朝采取崇佛政策，因而茶在这一时期得到极大的推广，饮茶得以普及，茶礼也初步形成。朝鲜时期（1392—1910）在崇儒抑佛以及酒风盛行②的大环境下，茶文化一度衰落。后在以草衣禅师（1786—1866）为代表的多位茶学大家的努力下，茶文化迎来了短暂的中兴。

Tea was first introduced to the Korean peninsula by monks in the 7th century. At that time, it was used on important rituals as ceremonial offerings. During the period of Unified Silla (668–935), people began to plant tea; and the way of processing and consuming tea was influenced by the neighboring Tang Court. In that period, tea was not affordable for common people, and it was consumed mainly by the upper class or in Buddhist temples. In the Koryo Dynasty (935–1392), thanks to the close connection of tea with Buddhism, tea got popularized as Buddhism prevailed. The boiled tea and then the whipped tea were introduced to the island. In the 15th century, Buddhist ceremonies were replaced by Confucian rites and wine replaced tea in formal ceremonies. So the custom of tea drinking long associated with Buddhism became a monastic custom. In the first half of the 19th century, with the efforts of many tea scholars represented by Chouiseonsa, Korean tea culture experienced a temporary revival.

朝鲜半岛进入日本殖民统治时期（1910—1945）后，茶文化再度

---

① 这一时期的茶礼分为官府茶礼、禅宗茶礼和儒教茶礼三种。
② 1416 年，法令规定在祭祀等活动中用酒或甜酒代替茶。

韩国宝城茶园（视觉中国　供图）

受挫。后来，随着西方文化的影响，半岛南部的韩国形成了以咖啡和红茶为中心的茶馆文化。20世纪七八十年代，作为韩国政府民族文化传统复兴的一部分，振兴绿茶文化活动使得民众开始关注绿茶。

During the period when the Korean Peninsula was a colony of Japan (1910–1945), tea culture suffered a serious setback. With the influence of American and European culture, people in South Korea mainly preferred coffee and black tea. In the 1970s and 1980s, the government successfully promoted the tradition of drinking green tea by advocating the revitalization of the traditional culture.

9.3 英国 / Britain

欧洲对茶的了解和利用远远晚于亚洲国家。据说欧洲关于茶的最早记录是源自16世纪的一位阿拉伯[①]旅行家。在大航海时代[②]，欧洲人开始对东方有了更多的了解。葡萄牙人首先发现了茶这种令人愉悦的饮品，当时旅居亚洲的传教士和商人将茶叶作为礼物带回葡萄牙。而大规模的"东茶西传"却始于荷兰人，他们最先将茶叶作为商品运到欧洲。1610年，荷兰东印度公司将第一批茶叶运到欧洲[③]。至此，欧洲人开始饮茶，大规模的茶贸易徐徐拉开帷幕[④]。

Since Europe is far away from China, people there got to know the pleasant beverage of tea much later than people in Asian countries. The earliest record of tea in Europe is said to be found in the statement of an Arabian traveler. It was in the period of the great discoveries that the European people began to know more about the Orient. It was the Portuguese who first discovered the pleasures of tea after missionaries and merchants who lived in Asia brought tea back home as a gift. Yet it was the Dutch who saw the commercial potential of this remarkable leaf. In 1610, ships of the Dutch East India Company brought the first batch of tea into Europe. In the following years, Europe's exposure and interests to tea continued to grow.

---

① 约9世纪中叶，有关中国茶叶的记载出现在阿拉伯人的著述中。

② 大航海时代，亦称地理大发现，指15世纪至17世纪，世界各地，尤其是欧洲发起的广泛跨洋活动与地理学上的重大突破。

③ 1607年，荷兰人将茶叶从中国澳门运至印尼的万丹，将茶叶列入贸易项目；1610年，将茶叶经爪哇运至荷兰；1616年，将茶叶运销丹麦；1657年，将茶叶运销法国。

④ 1669年，英国东印度公司将茶运到英国。17世纪晚期，英国东印度公司开始主导茶叶贸易。

在茶叶传到欧洲的最初几十年里，绝大部分地区仍将茶叶作为药用品，在药店出售。1662年，英国国王查理二世[①]与葡萄牙[②]公主凯瑟琳结婚。凯瑟琳公主出嫁时把茶叶作为嫁妆带到了英国，并用茶招待王室贵族。得益于王后的示范引领，茶很快受到王室贵族的追捧，人们爱上了茶的清新滋味或者说爱上了奇妙的东方滋味。在很长一段时间内，茶叶极其昂贵，只有上流社会才能消费得起[③]。

In the decades after tea was first transported to Europe, tea was mainly consumed as a medicine and sold in pharmacies. In 1662, King Charles II of England married Catherine, Princess of Portugal, who brought a box of Chinese tea to England. When she served Chinese tea as a pleasant beverage for ladies at court, she set a new trend of drinking tea in England. And soon people fell in love with the fresh flavor or the mysterious oriental taste. However, at that time, tea was incredibly expensive and only the upper class could afford such a delicacy.

1657年，第一批作为饮料的茶叶从伦敦的卡洛韦咖啡馆售出。一年以后，世界第一则茶叶广告出现在报纸上。18世纪初，第一家名为"川宁"的茶叶店在伦敦开门迎客[④]。19世纪晚期，饮茶在英国各阶层普及。由此，茶成为英国人最喜欢的日常饮料。

---

① 查理二世在克伦威尔统治期间流亡荷兰，而荷兰的上流社会早在几十年前就已经开始饮茶。由此认为，查理二世回到英国时，已养成饮茶的习惯。

② 葡萄牙是最早进行大航海的国家之一，所以葡萄牙人很早就接触到中国茶叶，并养成饮茶的习惯。

③ 欧洲从中国进口茶叶，距离远、时间长、运费高、危险大，故量少价高。1660年，英国茶叶进口量仅为226千克。茶价昂贵的另一个原因是税收。1784年，茶叶的税收从119%下降到12%，随之带来的是茶叶消费量的增加。即便如此，18世纪末，茶叶价格依然昂贵。

④ 作为英国历史最悠久的茶叶品牌，"川宁"是目前全球最知名的茶叶品牌之一，有着300多年历史的品牌首店今天依然生意兴隆。

The first time tea sold as a pleasant beverage was in Garraway coffeehouse in London in 1657. A year later, the first advertisement on tea appeared on a newspaper. In the early 18th century, the first tea shop named Twinings opened and soon tea shops began to appear throughout the country. By the second half of the 19th century, its popularity had spread to all corners of the country. Thus, tea became the favorite beverage of the British.

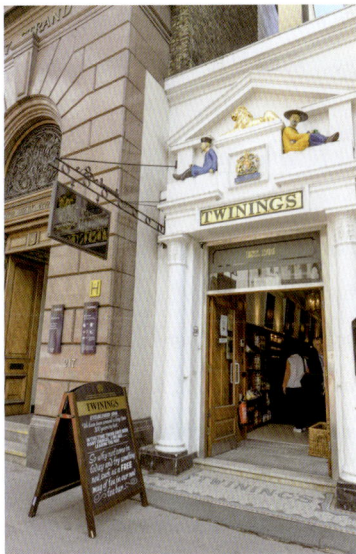

百年川宁老店（视觉中国　供图）

茶在英国得以普及的原因主要有两点：一是供给充足，随着英属殖民地国家种植的茶叶大量进入英国市场，茶叶价格下降；二是营销成功，茶被包装成一种保健效果极佳的功能饮料。

One important reason for the popularity of tea in Britain was its availability. With a large amount of tea supplied to the country by colonial countries, the price naturally went down. Besides, tea was labeled as a medicinal drink with wonderful "results"—a successful advertising and marketing strategy.

作为曾经的日不落帝国，英国通过各种方式将茶引入很多国家。19世纪上半叶之前，英国人想要喝茶只能从中国购买，导致大量白银流入中国。随着国内对茶叶需求的不断增加，英国不满足依赖进口中国茶，开始尝试在英国种植茶叶，但因土壤、日照、气候等方面的限制未能成功。为掌握茶叶贸易和消费的主动权，英国自 19 世纪下半叶起，陆续在其亚洲和非洲的殖民地尝试种植、生产茶叶，其中最成功的莫过于印度、斯里兰卡和肯尼亚。从这个意义上说，茶叶的全球商业种植在一定程度上得益于英国的推动。

Historically known as "the empire on which the sun never sets", Britain has played an important role in introducing tea to many countries. Before the first half of the 19th century, China monopolized the tea supply and the British spent a huge amount of money on tea, resulting in big trade deficits. With the increasing demand for tea, Britain decided to change the unfavorable trading pattern. They tried to cultivate tea in Britain, but failed because of unfavorable "terroir" [1]—the combination of soil, exposure and climate. In order to seize the initiative on tea trade and consumption, in the second half of the 19th century, Britain managed to plant tea and set up tea plantations in its Asian and African colonies, among which India, Sri Lanka and Kenya were the most successful. To some extent, the honor of global commercial tea cultivation goes to the British.

茶叶海外种植的成功后，英国还需将产自其殖民地的茶叶推销给市场。一方面，他们用茶叶拼配[2]的方法来弱化茶叶的产地概念；另一方面他们研制出一套茶叶的感官评审[3]标准，即凭借一款茶的色、香、

①　Terroir 意为风土，源于法语中的土壤一词 terre，原指在葡萄生长时，地理、气候因素、葡萄栽培法等影响红酒滋味和香气的因素总和，包括土壤条件、降水量、光照条件、风力、灌溉、排水等。
②　产茶殖民地负责茶叶种植和初制工序，茶叶拼配则在英国本土完成。今天，爱"拼"才会赢是茶界的共识，拼配技术实则取长补短，是各大品牌茶企的核心技术。
③　茶叶感官评审，俗称为"评茶"和"看茶"，指以人的感觉器官，如用眼看、口尝、鼻嗅、手摸来鉴定茶叶品质优劣、好坏的一种检验方法。目前世界上产茶国和消费国都是以茶叶感官评审方法来确定茶叶品质、级别和价格的。该方法简便，检验快速、结果准确，具有很强的适用性、通用性和广泛性。

味、型来评定茶叶好坏。在这两大法宝的加持下，产自印度、锡兰等国的茶叶迅速取代中国茶叶，成为国际茶业市场的主流，英国如愿成为红茶帝国。

After tea was successfully cultivated in colonial territories, the British had to face another tough challenge, that is, selling the tea produced in these colonies to the world. The first step the British took was to weaken the concept of Chinese tea. By blending Chinese tea with Indian tea and Ceylon tea, they produced a new product with mixed origin. The second step was to design a set of criteria which were used to evaluate objectively the quality of each variety, including the color, aroma, taste and appearance. Tea blending technology, together with tea evaluation system, changed the whole picture. Soon tea produced in India and Ceylon took the place of Chinese tea to be the most popular varieties in international tea market and Britain finally became the empire of black tea.

茶叶感官评审（静观　供图）

9.4 印度 / India

1824 年，在印度阿萨姆发现野生茶树。茶树或许很早就生长在印度，但印度对茶的栽培和饮用却源于英国人。1849 年，受英国东印度公司委派，苏格兰植物学家罗伯特·福琼（Robert Fortune）将从中国盗采来的茶树、茶种带至印度[①]，同时还带去了 8 名来自福建的制茶师傅。在中国制茶方法的基础上，英国人进行了改进，在实现生产标准化和机械化后，阿萨姆开始大量生产茶叶。阿萨姆的茶叶以稳定的品质及低廉的价格站稳脚跟，在国际贸易中逐渐取代中国茶叶。1856 年前后，英国开始在大吉岭开辟茶园，大吉岭红茶素有"红茶中的香槟"之称，其就源于福琼盗采的优良树种——来自中国福建的茶树。有了优良的树种和制茶技艺，加之茶园所在地区海拔和气候十分适宜茶树的生长，印度很快成为新的茶叶生产大国。

In 1824, wild tea plants were discovered in Assam. Tea may have grown in India long time ago, but the cultivation of tea and the custom of tea drinking in India were brought by the British. In 1849, the British East India Company sent Scottish botanist Robert Fortune to steal the crop from China and bring it back to British plantations in India. Meanwhile, he brought eight tea technicians from Fujian Province to India. In about 1856, tea plantations were set up by the British in Darjeeling. Today, Darjeeling black tea is known as the "champagne of black tea", and its tree species originated from the seeds stolen by Robert Fortune from Fujian province, China. With excellent tree species and tea techniques, coupled with the favorable altitude

---

[①] 关于这一段历史，可阅读《茶叶大盗：改变世界史的中国茶叶》（*For All the Tea in China: How England Stole the World's Favorite Drink and Changed History*）。

大吉岭茶园（视觉中国　供图）

and climate for the growth of tea, tea industry developed incredibly quickly. India soon became a major tea producer and exporter. The altitude and climate of northern India are very suitable for the growth of tea. As a result, tea plays an important role in economy and soon India became an important exporter of black tea.

今天，得益于适合茶叶生长的优良的风土，以及印度对技术改良的重视、不断的创新和战略性的市场推广等一系列措施，印度茶叶的品质得到全球公认，印度也成为仅次于中国的重要产茶大国 [1]。

Today, owing to strong geographical indications, heavy investment in

---

[1]　1933年至2005年，印度是世界最大的茶叶生产国。2006年之后，中国的茶叶产量超过印度，保持世界第一。近几年，中国茶叶产量约是印度的 2 倍，茶园面积是印度的 5 倍。

tea processing units, continuous innovation, and strategic market expansion, India is the world's second-largest tea producer only after China and Indian tea is one of the finest in the world.

9.5 斯里兰卡 / Sri Lanka

19 世纪下半叶，英国人将斯里兰卡带上茶叶种植的道路，将饱受咖啡叶锈病之苦的咖啡树替换成茶树。有"锡兰茶之父"美称的苏格兰人詹姆斯·泰勒[①] 于 1867 年开辟了第一个锡兰商业茶园，为斯里兰卡一个多世纪以来最大的出口产业奠定了基础。1873 年，詹姆斯·泰勒生产的 23 磅茶叶运至伦敦，锡兰茶首次在国际上亮相。

Tea was brought to Sri Lanka by the British in the second half of the 19th century, when the coffee plantations, the foundation of the economy, were wiped out due to the coffee blight. Known as "father of Ceylon tea", James Taylor, Scottish by origin, started the first commercial tea plantation in the island in 1867, laying the foundation for what would become Sri Lanka's largest export industry for over a century. In 1873, Ceylon Tea made its international debut when twenty-three pounds of tea produced by James Taylor reached London.

19 世纪 80 年代，斯里兰卡茶业快速发展。到 80 年代末，因为茶业可带来更大的利润，几乎所有的咖啡种植园都被改造成了茶叶种植园。锡兰的茶业能如此快速发展，全球最大的茶叶品牌"立顿"的创

---

① 詹姆斯·泰勒的传奇始于一场几乎摧毁了全岛 50 万英亩咖啡产业的枯叶病。庆幸的是，泰勒在这场危机中看到转机。他注意到他试验性种植的茶树并未受到枯叶病的影响。这最初的 19 英亩茶园，最终为数以百万计的斯里兰卡人解决了生计问题。

始人托马斯·立顿爵士功不可没。他投资锡兰红茶，并于 1890 年在英国推出立顿红茶，广告词是"从茶园直接进入茶壶的好茶"。

During the 1880s, tea production in Sri Lanka grew rapidly. By the late 1880s, almost all the coffee plantations had been converted to tea as it was seen as a more lucrative alternative. Sir Thomas Lipton, the founder of Lipton, the world's largest tea brand, contributed to the rapid development of Ceylon's tea industry. He bought tea plantations in Ceylon and by telling customers that his tea was "Direct from tea garden to the tea pot", he successfully launched "Lipton" in Britain in 1890.

1965 年，斯里兰卡成为世界上最大的茶叶出口国。

In 1965, Sri Lanka became the largest exporter of tea in the world.

斯里兰卡地处热带，年均气温 28℃、干湿分明、昼夜温差大，加之肥沃的土地，为茶叶生长提供了得天独厚的自然条件，所有这些，

斯里兰卡采茶人（视觉中国　供图）

造就了斯里兰卡红茶的绝佳品质。今天，斯里兰卡是全球主要的茶叶生产国和出口国，为超过 100 万的斯里兰卡人（约占全国人口数量的5%）解决了生计问题。

Located in the tropics, Sri Lanka enjoys unique natural conditions for tea growth with an average annual temperature of 28℃, distinct wet and dry seasons, large temperature difference between day and night, and fertile land. So tea produced in Sri Lanka is one of the best in the world. Today, Sri Lanka has become one of the largest tea producers in the world and also one of the greatest exporters of teas. Over a million of the Lankan workforce—that's approximately 5% of the total population—are involved in the tea industry.

9.6 肯尼亚 / Kenya

与大多数国家一样，非洲的茶最初也源于中国。据记载，早在 14 世纪摩洛哥和索马里学者曾讨论中国的茶习俗。此外，中国也曾将茶作为贵重的礼物赠送给非洲的领袖。17 世纪初，葡萄牙和荷兰商人途经非洲时，再一次将茶及其文化带到非洲。19 世纪末，英国开始在非洲殖民地推广茶叶种植。今天，绝大部分非洲国家都种植茶叶，肯尼亚等国家还成为世界主要茶叶出口国。

Similar to most countries in the world, tea was first introduced to Africa from China. It is recorded that in the 14th century, some Moroccan and Somali scholars discussed Chinese tea customs; and tea was once presented to African leaders as a gift. In the early 17th century, when the Portuguese and Dutch traders traveled across Africa, they brought the custom of tea drinking to some African countries. In the late 19th century, tea was cultivated in British colonial territories in Africa. Today, many African countries grow tea, and some of them such as Kenya have grown into major tea producers and exporters.

1903 年，英国人将茶叶引入肯尼亚。1920 年，英国殖民肯尼亚后，开始大规模种植茶叶。1963 年肯尼亚宣布独立，但英国人留下的茶叶种植技术和管理理念一直沿用至今。

Tea was introduced to Kenya by the British in 1903, and in 1920, large-scale cultivation of tea started in Kenya. After the independence of the country in 1963, the pattern of tea cultivation and processing remained unchanged.

肯尼亚自然气候和地理环境优越、茶树种植全程无污染、采制精细、价格合理，故颇受消费者欢迎。肯尼亚已成为全球重要的茶叶生产国和出口国之一。肯尼亚主产 CTC 红茶，这样的产业布局既有历史原因也是为茶叶出口服务，因为红茶是世界流通最广的茶类。

Kenyan tea is popular with international consumers for several reasons. First, due to its favorable climate and soil, Kenyan tea contains no artificial chemicals or pesticides. Besides, Kenyan tea is known for its fine manufacturing technology and reasonable price. Kenya has long been a major tea producers and exporters. The country mainly produces black tea and CTC teas dominate. It is the tradition since its colonial days; also it caters to the international market.

茶业在肯尼亚经济中占有重要地位，肯尼亚茶叶以外销为主，是肯尼亚主要创汇产业之一。10% 的肯尼亚人都直接或间接从事茶业。

The tea industry is of great importance in Kenya. It is one of the biggest

肯尼亚采茶人（视觉中国　供图）

export earners of the country[1] and one out of ten Kenyans is employed in tea or tea-related industries.

9.7 俄罗斯 / Russia

俄罗斯对茶的记载始于 1567 年[2]，而俄罗斯人的喝茶历史最早可追溯到 1618 年，当时中国明朝派使节携带茶叶作为礼品赠送给沙皇，由此开创了俄国人品饮中国茶的先河。1638 年，使臣带回了由蒙古可

---

[1]　Tea has contributed over 20 percent of total foreign exchange earnings in Kenya for many years.

[2]　1567年，俄国人彼得洛夫和雅里谢夫向本国介绍报道茶树的新闻，为俄国茶事记载之开端。

汗赠送给沙皇的中国茶叶。17 世纪，中俄之间由政府主导的商贸通道正式形成[1]，因该通道主要承担茶叶贸易与运输而被称为"茶叶之路"，又称"万里茶道"。彼时，茶叶主要通过骆驼商队运输，时间长、成本高，因此只有王公贵族才能喝得起茶。19 世纪末，俄罗斯从中国引进茶树试种成功。1893 年，刘峻周[2] 等十名中国茶叶技工抵达俄罗斯格鲁吉亚[3] 传授制茶技艺，所产制的茶叶在 1900 年巴黎世界博览会上获金奖。20 世纪初，随着西伯利亚铁路的开通，茶叶价格大幅降低，饮茶逐渐成为俄罗斯民众生活的一部分[4]。

The earliest record of tea in Russia was in 1567. And the history of tea drinking in Russia began in 1618, when tea was sent to the Tsar by the Ming Court. In 1638, the envoy brought back Chinese tea gifted to the Tsar by the Mongol Khan. In the 17th century, the China-Russia Tea Road of Ten Thousand Miles across the Eurasian continent began to take shape. It was about regular tea supplies from China via camel caravan. However, the difficult trade route made the cost of tea extremely high, so the beverage became available only to royalty and the very wealthy of Russia. At the end of 19th century, Chinese tea was successfully planted in Russia. In 1893, to improve quality of the tea produced in Russia, Liu Junzhou, together with nine Chinese tea technicians, was invited to Georgia, Russia. Thanks to their efforts, the tea made by them won the gold medal at the 1900 Paris World

---

[1]　1679年，中俄签订了第一笔购茶合同。当时显贵们饮茶看重的是茶的提神功效，而非将其作为可口的饮品。

[2]　因对俄国茶业发展做出杰出贡献，刘峻周被誉为"中国茶王"。1924年，刘峻周被授予"劳动红旗"勋章，其旧居被设为"茶叶博物馆"。

[3]　19世纪初，格鲁吉亚被沙皇俄国兼并。1991年4月9日，格鲁吉亚共和国宣布独立；1995年8月24日，定国名为格鲁吉亚。

[4]　以骆驼商队形式运送茶叶至莫斯科需要一年以上，而铁路运输仅需一周左右。交通的便利，缩短了运输时间、降低了成本，传统的中俄万里茶道逐渐消失。

Expo. At the beginning of the 20th century, the Trans-Siberian Railway greatly drove down the the price of tea and gradually tea became a daily beverage for Russian families.

9.8 土耳其 / Türkiye^①

　　土耳其位于地中海和黑海之间，横跨欧亚两大洲，有"东西桥梁"之称。土耳其的前身是奥斯曼帝国^②，国土横跨欧亚非，控制着东西方贸易路线。早在帝国时期，茶叶通过丝绸之路就来到了这片土地。19世纪末，奥斯曼帝国从中国引进了茶树，开始试种茶叶，不过没有成功。1923 年，土耳其共和国成立后不久，土耳其从格鲁吉亚^③引进茶籽，开始生产茶叶。在政策的持续推动下，20 世纪 60 年代，土耳其茶产业发展迅速，基本可以自给自足。今天，土耳其茶叶产量和消费量均居世界前列。

Nestled between the Black Sea and the Mediterranean Sea, Türkiye is partly in southwest Asia and partly in southeast Europe. It is where the East meets the West. Formerly known as Ottoman Empire, its territories extended to Europe, Asia and Africa at the height of its power. Tea was introduced to the empire through the Silk Road in the 16th century. At the end of the 19th century, Chinese tea plants were introduced to the land but the experiment of cultivating tea plants failed. Shortly after the founding

---

　　① 2022年5月，土耳其官方外文名从Turkey正式更改为Türkiye，理由是"更能体现土耳其民族的文化、文明和价值观"。旧国名Turkey，亦指"火鸡"，故土耳其在国际社交中常被调侃为"火鸡共和国"。此外，Turkey 还表示"失败的事情"，或"蠢货"。

　　② 奥斯曼帝国（1299—1922），是土耳其人建立的多民族帝国，位处东西文明交汇处，掌控东西文明的陆上交通线达 6 个世纪之久。

　　③ 彼时，格鲁吉亚是苏联的一部分。

土耳其采茶人（视觉中国 供图）

of the Republic of Türkiye in 1923, with tea seeds from Georgia, Türkiye restarted the experiment of tea cultivating. Thanks to incentive policies of the government, by the 1960s, Türkiye had been totally self-sufficient in tea production. Today, both tea production and consumption of Türkiye are among the highest in the world.

9.9 美国 / The United States

17 世纪，荷兰殖民者首次将茶引入美国。英国人控制北美后，英式茶饮习俗在当时的殖民地费城、波士顿等地迅速普及。1773 年 12 月 16 日，为抗议所谓的《茶法》，即授予英国东印度公司在美国销售茶叶的垄断权，并限制了殖民地人们从其他来源购买茶叶的自由，自由之子示威者将停靠在波士顿港的英国东印度公司商船上的茶叶倒进大海。这就是著名的波士顿倾茶事件，也是美国独立战争的导火索。经由此事，美国人一度抵制象征殖民的茶叶，转而消费咖啡等饮品。

Tea was first introduced to America by the Dutch settlers during the 17th century. After the British took over the region, the English tea culture and customs spread fast in colonies like Philadelphia and Boston. However, the British Parliament passed the Tea Act, which gave the British East India Company a monopoly on the sale of tea in America and removed the colonists' ability to purchase tea from other sources. Frustrated and angry at British unfair policies, colonists boycotted the Britain's goods and the Sons of the Liberty rushed to the East India Company's ship and dumped all the tea to the sea on December 16, 1773. This is the famous Boston Tea Party, which sparked off the American War of Independence. It eventually led to the United States of America becoming an independent nation instead of a group of British colonies. After this political event, tea was not popular as it used to be among Americans and some of them switched to coffee and other drinks.

美国超市琳琅满目的茶

结语 / Conclusion

除了上述国家外，早在 17 世纪，茶叶也通过葡萄牙和西班牙殖民者传到了南美洲。直到 20 世纪初，南美洲部分国家才开始茶叶栽培。1924 年，阿根廷开始种植中国茶种；如今，阿根廷主产红茶，已成为南美第一大、全球十大茶叶生产国之一。

In addition to the above-mentioned countries, tea was also introduced to South America by the Portuguese and Spanish as early as the 17th century. Tea farming in the South America started in the early 20th century. A Chinese tea species was first planted in Argentina in 1924. Presently, Argentina is the largest tea producer in South America and one of the TOP 10 tea producers in the world. Black tea is the primary variety of tea produced in this country.

可以说，今时今日，茶已飘香于世界的各大个角落。

Today, the fragrance of tea has spread to every corner of the world.

茶是中华文化的重要组成部分，也是中国人热情好客的象征。随着中国茶及其文化在海外落地生根、萌发新芽，异域茶风又吹回中国，为中国茶文化增添新彩。今天，除了本土的茶外，国内消费者还可以品尝到斯里兰卡的红茶、印度的拉茶、日本的抹茶等来自世界各地的特色茶。除此之外，还可以通过一场惬意的英式下午茶，品味异域文化之优雅。

Tea is an important part of Chinese culture and a symbol of hospitality of Chinese. As Chinese tea and the culture has spread to all over the world, it has also been enriched by other civilizations. Today, in addition to Chinese tea, a variety of exotic teas are available to Chinese customers such as black tea from Sri Lanka, Masala Tea from India, and Matcha from Japan. In addition, people can easily have a glimpse of English culture by spending a couple of agreeable hours chatting over a cup of English afternoon tea.

茶是承载中国历史和文化的名片，以茶为媒，各美其美、美美与共。正如习近平主席在 2020 年 5 月 21 日首个"国际茶日"的贺信中所言："作为茶叶生产和消费大国，中国愿同各方一道，推动全球茶产业持续健康发展，深化茶文化交融互鉴，让更多的人知茶、爱茶，共品茶香茶韵，共享美好生活。"

Serving as a medium for cultural communication and mutual learning across the world, tea has become China's "business card", telling stories of this ancient and modern country. As President Xi Jinping said in his congratulatory letter on May 21, 2020, the first "International Tea Day", "China, a major producer and consumer of tea, is willing to work with all sides to nurture the sustained and healthy development of the global tea industry, deepen cultural exchanges on tea, and allow more people to relish lives accompanied by tea."

小贴士：

1. 英国民众的茶叶消费受国际国内大环境影响，也曾有起伏。如在"二战"时期，从 1940 年 7 月开始，为了更好地管理红茶，英国政府实行了红茶配给制；战争结束后，为了恢复破败的经济，更好地管理外汇、及时偿还战争负债，继续实行了红茶配给制，直至 1952 年 10 月结束。20 世纪 60~70 年代，流行的美国快餐和咖啡使茶失去了曾经的重要性；80 年代后，尤其是 2000 年后，茶再次获得英国民众的关注和喜爱。鉴于英国在茶叶全球化过程中涉及内容很多，了解以上内容即可。

2. 在讲解茶的全球化时，可结合当时英国的殖民地如印度、斯里兰卡、肯尼亚等国茶叶种植历史。1600 年，英国入侵印度；1757 年印度沦为英殖民地；1849 年全境被英占领。1947 年 6 月，英国通过"蒙巴顿方案"，将印度分为印度和巴基斯坦两个自治领。1947 年 8 月 15 日，印度独立。1950 年 1 月 26 日，印度成立共和国，同时仍为英联邦成员。斯里兰卡 16 世纪起，先后被葡萄牙人和荷兰人统治；1802 年正式成为英国殖民地。1948 年 2 月获得独立，定国名锡兰。1972 年 5 月 22 日改称斯里兰卡共和国。1978 年 8 月 16 日改国名为斯里兰卡民主社会主义共和国。1890 年，英、德瓜分东非，肯尼亚被划归英国，英政府于 1895 年宣布肯尼亚为其"东非保护地"，1920 年改为殖民地。1964 年 12 月 12 日，肯尼亚共和国成立，但仍留在英联邦内。

　　"凡是有中国人的地方就有茶"[1]，茶作为全球最大众化、最受欢迎、最有益于身心健康的饮品已成为中华民族的举国之饮。数千年来，茶被赋予很多意义，如表示欢迎、尊敬、歉意、祝福、接纳等；同时，茶在提升人的道德修养、人格完善方面发挥着重要的作用；此外，茶还是社交领域以及婚礼、拜师、祭祀等仪式的重要组成部分

　　"Where there are the Chinese, there is tea"[2]. As a popular and healthy beverage in the world, tea has long been the national drink of China. It is more a cultural statement than a beverage. Tea holds many meanings in China. Tea can be used to convey hospitality, respect, apology, forgiveness,

---

[1]　此句源自梁实秋的散文《喝茶》，原文如下："茶是我们中国人的饮料，口干解渴，惟茶是尚。茶字，形近于荼，声近于槚，来源甚古，流传海外，凡是有中国人的地方就有茶。"

[2]　该句也可表述为 "Wherever the Chinese go, the custom of drinking tea follows"。

blessing, and sharing moments together. Meanwhile, tea has played an important role in improving moral cultivation and perfection of personality. Besides, tea is also an important medium for communication in socializing and ceremonies such as weddings, apprentice-taking and sacrifices and harmonious interpersonal relations.

大俗之茶（静观　供图）

10.1 中国常见茶俗 / Common Tea Customs in China

中国民间有许多与茶相关的风俗和礼仪，这些茶礼的核心都是一个"敬"字。最常见的习俗是以茶待客，也就是"客来敬茶"①。奉上一杯茶，不仅表达了对客人的欢迎和认同，也营造了轻松愉悦的氛围。

---

① 以茶待客的习俗至少在魏晋南北朝时已经出现。吴兴太守陆纳待客，"所设唯茶果而已"。唐宋时期，"客来敬茶"已成普遍习俗。宋代杜耒的"寒夜客来茶当酒，竹炉汤沸火初红"更是生动表现了这一习俗。

Tea is deeply rooted in Chinese customs and traditions. The most common one is greeting guests with a cup of tea; in other words, tea is served immediately after guests taking their seats in a Chinese family. Serving a cup of tea is a sign of hospitality; it also provides a sense of identity and creates a relaxing and pleasant atmosphere.

茶可迎客，也可送客。"端茶送客"是一种委婉的逐客令，主人不用一语，客人心领神会，既避免了主客的尴尬，也顾及了双方的颜面。这一习俗在清代及民国时期较为常见。

Tea could also be used to drop a hint to guests to end the conversation and say goodbye, which was a common practice during the Qing Dynasty and the Republican Period. When the host held the Gaiwan, it was a euphemistic way to tell the guest that it was time for him/her to leave. Taking in the hint, the guest would end the conversation and took the leave. In this way, the visit was naturally finished without even a word.

待客敬茶时，一般不会倒满，因为茶满容易溢出；且茶水通常比较烫，即使不溢出也不便客人端杯品饮。中国人对此的解释是："茶倒七分满，余下三分是情谊。"这个看似简单的日常之举，反映的却是中国人"做人做事需留有余地"的生活智慧。

When pouring tea, Chinese people usually fill the cup 70% full, leaving 30% empty as a gesture of friendly sentiments. A more reasonable explanation for this is that newly brewed tea is very hot, so it is more convenient for guests to hold the cup that is 30% empty. For Chinese, such a practice is more than a matter of safety. It reflects Chinese traditional wisdom that whatever we are doing, we should always leave some leeway for future possibility.

此外，饮茶时还应遵守一些基本的社交礼仪，最基本的就是先尊后卑，先老后少，即等长辈、主客、上级开始喝茶后，其他人再举杯品饮；斟茶时，一般也是晚辈为长辈、下级为上级服务。看似简单的礼数，彰显着中国人谦和礼敬的人文精神。

Besides, the general social propriety should be observed when drinking tea. For example, it is the young or the junior employee that serves tea for the elder or the senior; only after the elder, the main guest and the superior start drinking tea, can others raise their tea cups. It shows the humanistic spirit of modesty and courtesy of Chinese people.

素有"礼仪之邦"美称的中国，讲究尊师重道，拜师敬茶的仪式古已有之。在敬茶环节，徒弟须举杯齐眉，以腰为轴，躬身将茶献出，以示尊敬。

China has always been known as a land of propriety and the Chinese have the utmost respect for teachers. Since ancient times, tea has been used on the ceremony of apprentice-taking. Apprentices are expected to serve tea for masters or teachers by holding the cup high to the level of their eyebrow and making a deep bow. In this way, they show their greatest respect to teachers.

茶还可以表示歉意，致歉一方泡茶、敬茶表达歉意，如果对方接受敬茶，则表示接受道歉并原谅致歉人。正因如此，在中国，茶馆也成为解决纠纷、重修旧好的一个场所。

In addition, tea can be used to convey apology. Serving tea is a mild and roundabout way to express apology and if the tea is accepted, so is the apology. So Chinese teahouses is a place to resolve disputes and patch up relations.

江南和川渝民间曾流行"吃讲茶"的习俗①，即在茶馆进行民间纠纷的裁决。纠纷双方相约茶馆，人手一杯茶，纠纷双方或请位重之人调停，或请在场茶客评判，最终大家在茶馆中或握手言和，或不欢而散。茶馆氛围轻松、随意；茶性平和，使人冷静。以"吃讲茶"的方式解决民间纠纷，形式灵活、成本低廉，既不伤情面又可化解矛盾。"吃讲茶"这一习俗所营造和利用的正是茶"和为贵"的内涵②。

In Jiangnan③ and Sichuan-Chongqing area, the custom of "Chi Jiang Cha" used to be popular. The so-called "Chi Jiang Cha" means "resolving a dispute in a teahouse". The two parties will meet at a teahouse, inviting a person with high prestige as the meditator. After the two parties' stating and arguing, the mediator will give a verdict, on which the two parties can either agree or disagree. If the decision is accepted by the two parties, the negotiation will end with a satisfying cup of tea. "Chi Jiang Cha" is a flexible way to settle disputes at low cost, which helps solve the problem in a fairly friendly atmosphere. As the teahouse provides a harmonious environment and tea helps cool people down, the custom of "Chi Jiang Cha" reflects the connotation of tea—"Harmony is most precious".

中国还有以茶祭祀的习俗。这一习俗大致是在两晋之后才逐渐兴起的④。我国的祭礼习俗中，历来流传以茶、酒、水果、鲜花等作为主

---

① "吃讲茶"在清代至民国时期非常流行，新中国成立后逐渐淡出民间。2006年6月，"吃讲茶"被列入绍兴市非物质文化遗产名录。

② 如遇评判不公、一方不服，或是一方蛮不讲理，也会出现双方大打出手的情况，故"吃讲茶"存在治安隐患，旧时不少茶馆悬挂标牌禁止讲茶。

③ Jiangnan, which literally means south of the Yangtze River, refers to the region around the lower reaches of the River.

④ 以茶为祭的正式记载最早出现于《南齐书·武帝本纪》："我灵上慎勿以牲为祭，唯设饼、茶饮、干饭、酒脯而已……"

庙会（赶茶场）

要祭品。用茶作祭，一般有三种形式，一是干茶，二是冲泡好的茶汤，三是以茶壶、茶盅象征茶。茶之所以被用作祭品，是因为古人认为茶乃洁净之物。今天，祭祀的形式已发生很大的变化，但以茶为祭的习俗依旧[①]，尤以云南少数民族为甚。

    In addition, tea has long been used as a sacrifice, which is thought to have originated in the Jin Dynasty (265–420). Along with liquor, fruits and flowers, tea was offered as a sacrifice to gods and ancestors, since the ancients believed tea was pure and clean. Generally, tea is used on such occasions in three different ways: tea leaves, tea liquid, and tea utensils representing tea. Today, in spite of great changes in the form of sacrifice, tea is still an important part on such occasions, especially for ethnic minorities in Yunnan province.

---

    ① 今天，云南的布朗族、布依族、彝族、傣族均有以茶为祭的风俗。2013年，"普洱祭茶祖习俗"入选云南省非物质文化遗产名录。

茶礼是我国古代婚礼的重要组成部分，这主要源于"茶不移本，植必子生"①的认知，即茶树只能从种子萌芽成株，一旦移植就会枯死。茶的这种品质与人们对婚姻"从一而终"的期盼不谋而合②。一般认为，茶叶与婚礼结缘始于唐朝。据说文成公主入藏时，嫁妆中便有茶叶，之后茶叶便成为出嫁时的必需品。到了宋代，"三茶六礼"③婚庆习俗形成，"三茶"指的是订婚时的"下茶"④、结婚时的"定茶"及同房时的"合茶"。我国许多地区和民族至今仍有"茶为媒"的风尚。今天婚礼中的茶礼既有对新人"执子之手，与子偕老"的祝愿，更是基于茶的特性"和"，表达对新人"凤协鸾和"的祝福。

In the past, tea was an indispensable part of marriage since tea was a symbol of faithfulness. It was believed that tea plants could only spout from seeds; and once transplanted, they would die. This quality or merit of tea coincided with people's expectation of marriage—"being faithful for life to marriage". It was in the Tang Dynasty that tea became part of the wedding ceremony. It is said that when Princess Wencheng went to Tubo (present day Xizang), she took tea as part of her dowry. Since then, tea has been an indispensable part of a woman's dowry. Since the Song Dynasty, the traditional marriage custom of "three teas and six rites" has been observed.

---

① 此句源于明代许次纾《茶疏》。该书是一部综合性的茶叶专著，成书于1597年。

② 清代福格在《听雨丛谈》中说到古人以茶为聘："今婚礼行聘，以茶叶为币，满汉之俗皆然，且非正室不用。近日八旗纳聘，虽不用茶，而必曰'下茶'，存其名也。"之所以如此，福格解释道："行聘以茶，盖种子既种，不能更移，与奠雁之义同。"

③ "六礼"指求婚至完婚全过程，即婚姻据以成立的纳采、问名、纳吉、纳征、请期、亲迎等六种仪式。

④ 男方婚娶下聘称为"下茶"，女方接受男方聘礼则称为"受茶"。《红楼梦》第二十五回"魇魔法叔嫂逢五鬼 通灵玉蒙蔽遇双真"中，王熙凤就曾借着吃茶之机打趣宝玉、黛玉。黛玉笑道："你们听听，这是吃了他们家一点子茶叶，就来使唤人了。"凤姐儿笑道："倒求你，你倒说这些闲话，吃茶吃水的。你既吃了我们家的茶，怎么还不给我们家作媳妇？"

The so called "three teas" refers to "sending tea as a token of love" at the time of engagement, "drinking a cup of auspicious tea" at the wedding ceremony, and "the couple drinking the same cup of tea symbolizing being together for a whole life" at the end of the ceremony. This tradition is still kept in many regions. Today, apart from the traditional view that marriage is meant to last for a whole life, people attach more beautiful expectation to marriage with tea — harmony and happiness of future life.

茶与婚礼

10.2 中国少数民族茶俗 / Tea Customs of China's Ethnic Minorities

中国很多民族都有与茶相关的习俗。汉族饮茶虽茶品不同、冲泡方法有别，却大都推崇清饮，认为清饮最能保持茶的纯粹、体会茶的本色。而很多少数民族同胞喜欢在茶中加入其他丰富的食材，创造出极具地域性、民族性的茶俗。

Many ethnic groups have their own tea-drinking customs. Generally, in spite of the regional preferences in tea and brewing methods, Han people prefer plain tea because they believe it is the best way to enjoy the fragrance and flavor of tea. While many ethnic minorities tend to mix tea with local ingredients, creating rich regional and ethnic tea flavors.

（1）白族"三道茶" / Three-course Tea of the Bai Nationality

白族"三道茶"是白族人民在传统婚礼、招待贵宾时的一种极具地方特色与民族风韵的饮茶习俗。它以其独特的"头苦、二甜、三回味"的特点类比人生进程：第一道"苦茶"，是用小陶罐烤大理特产沱茶[①]，茶叶烤黄烤香后，冲入滚烫开水。此道茶以浓酽为佳，口味清苦，代表人生的苦境。第二道"甜茶"，是以大理特产乳扇、核桃仁和红糖为佐料，冲入茶水制作而成。此道茶甜而不腻，寓意苦尽甘来。第三道"回味茶"，是茶水中加入蜂蜜、花椒、姜片、桂皮、炒米花等制

---

① 该环节又称烤茶或百抖茶（因烤茶时不断抖动陶罐，故名）。"三道茶"的口感首先取决于茶叶。常用之茶除了日常所用的苍山绿茶外，还有下关沱茶，该茶是一种制成圆锥窝头状的黑茶紧压茶。2011 年，"黑茶制作技艺（下关沱茶制作技艺）"入选国家级非物质文化遗产名录。

白族"三道茶"

作而成。此道茶甜麻香辣、回味无穷，恰如生活的百转千回。可以说，"三道茶"还原了人生最基本的滋味，是白族人民最朴素的人生经验与智慧①。

For the Bai nationality, the most respectful way of treating guests is to offer the three-course tea. It is also served on important occasions such as wedding ceremony. Three-course tea tastes bitter for the first brew, sweet for the second, rich and lasting for the third, which interestingly implies the vicissitude of life. The first course is made by baking a kind of compressed dark tea named Tuo tea, a local specialty of Dali, in a small pottery pot. When the tea turns yellow and gives off fragrance, pour some boiling water into the pot. The first course of tea is strong and tastes bitter, representing the hardships of life. The second course is made by adding milk fan, another

---

① "三道茶"的另一种解读是：第一道"苦茶"，寓意要立业先吃苦；第二道"甜茶"，有先苦后甜之意；第三道"回味茶"，寓意凡事需多"回味"。

local specialty, walnut kernel and brown sugar into the tea. So the tea is sweet, signifying that happiness comes after hardships. The third is made by adding honey, Sichuan pepper, ginger slices, cinnamon and fried rice into the pot. It tastes sweet and spicy, with rich aftertaste, just like the twists and turns of life. To some extent, the three-course tea resembles ups and downs of life and it is the wisdom of Bai nationality.

2014 年，"茶俗（白族'三道茶'）"入选国家级非物质文化遗产代表性项目名录。今天，"三道茶"与白族传统歌舞融合，衍生出白族"三道茶歌舞"，成为云南文旅的一张名片。

In 2014, "the three-course tea of Bai nationality" was included in the national intangible cultural heritage list. Today, locals have skillfully combined the three-course tea customs with their traditional dance, which has been a spokesman of cultural tourism of Yunnan province.

（2）藏族酥油茶 / Yak Butter Tea of the Tibetans

西藏素有"世界屋脊"之称，这里海拔高，气候寒冷干燥。当地百姓常年以奶、肉、青稞①为主食，因瓜果蔬菜很少，茶叶成为当地人民不可缺少的食品。藏族同胞喝茶，既有喝清茶、奶茶的，也有喝酥油②茶的，当然，喝得最多的还是酥油茶。酥油茶是一种在茶汤中加入

---

① 青稞常做成青稞炒面，即糌粑（音 zān ba）；糍粑（音 cí bā）是用糯米蒸熟捣烂后所制成的面食。此处应注意两者的区别，避免讲解混淆。

② 酥油是把牛奶或羊奶煮沸，搅拌后倒入竹筒内冷却后，凝结在液体表面的一层脂肪。在跟国外朋友介绍时，可简单表述为：酥油是类似黄油的一种乳制品。如此，外国人比较容易理解。

酥油、盐和糖等原料，经数道工序加工而成的茶①。制作酥油茶的茶叶，一般选用黑茶紧压茶②。

Known as the "roof of the world", Xizang is located at high altitudes, with cold and dry climate. Since the local diet consists largely of meats, milk products and barley, with very few vegetables and fruits, tea is a daily necessity. People there prepare tea in different ways with the most popular being yak butter tea which is made by mixing tea, yak butter, salt and sugar together. The process consists of several steps and the tea used is brick tea (a kind of compressed dark tea).

据说酥油茶源自文成公主。唐贞观十五年（641），文成公主下嫁吐蕃赞普松赞干布，公主的和亲队伍用了两年时间走完从长安到拉萨3000多千米的和亲之路。途中茶叶缓解了人们的高原反应。当和亲队伍抵达拉萨后，饮茶习俗也随之带入。据说公主为适应当地饮食习惯，将当地人常食用的酥油和食盐加入茶水中，在一碗酥油茶中实现了汉藏融合。酥油茶既可御寒、助消化，又可补充营养，除了日常饮用外，在一些重要场合也是必不可少的。在藏族人家，客来敬茶敬的就是酥油茶，代表着主人的好客；当孩子出生时，亲朋好友会带着酥油茶送来祝福；当有人远行时，也少不了一碗酥油茶，送上一路平安的祝福。

It is said that yak butter tea originated from Princess Wencheng in the Tang Dynasty, who married the king of Tubo, present day Xizang, in 641.

---

① 加工酥油茶一般先烧水，水沸后，用刀把紧压茶捣碎，放入沸水中，煮好后滤去茶渣，把茶汤倒入打茶筒内。同时，用另一口锅煮牛奶，煮到表面凝结一层酥油时，也倒入打茶桶内，再放入适量的盐和糖。然后盖住打茶桶，用长棍不断搅拌，待茶、酥油、盐、糖等充分混合，酥油茶就打好了。

② 以普洱茶、金尖为最常用茶。

茶具多穆壶

藏族酥油茶
（视觉中国　供图）

藏族酥油茶茶具——打茶桶

It is said that on her way to Tubo, the Princess took tea as part of her dowry. It took them two years to get to Lhasa from Chang'an (then the capital of the Tang Dynasty). And during the arduous 3000-kilometer-journey, tea helped ease the altitude sickness. When they finally arrived in Lhasa, so did the custom of tea drinking. Later the Princess combined the tea-drinking custom with the local diet by adding butter and salt into tea. Yak butter tea can keep out the cold, help digestion, and supply the needed nutrients. It is a perfect example of integration of different cultures and nationalities. As the most popular beverage among Tibetans, it is also served on important occasions. For example, it is offered to welcome guests; it is presented on the birth of a baby, representing best wishes to the coming of a new life; it is also prepared to wish people a safe journey.

（3）蒙古族咸奶茶 / Salty Milk Tea of the Mongolians

蒙古族茶习俗的形成大约始于 13 世纪初的民族大融合时期。蒙古族以食牛、羊肉及奶制品为主，茶是牧民不可或缺的饮品。咸奶茶除了可以解渴外，还可以补充营养。在牧区，人们习惯"一日三餐茶"。

Salty milk tea is the most popular beverage of Mongolians, which started from the early 13th century when Mongolians founded the Yuan Dynasty. Since the local diet consists largely of beef, mutton and milk products, with very few vegetables and fruits, tea is a daily necessity. Drinking salty milk tea can not only quench thirst, but also supply the needed nutrients. In pastoral areas, people drink salty milk tea at each meal.

蒙古家庭每天清晨的第一件事就是煮一锅咸奶茶，供全家整天享用。咸奶茶通常由砖茶煮成，加入鲜奶熬制，饮用时加盐，有时还会加入黄油、炒米、奶制品甚至肉类。煮咸奶茶看似简单，实则不然，滋味的好坏、营养成分的多少与煮茶时用的锅、放的茶、加的水、掺的奶、烧的时间以及先后次序都有关系，只有器、茶、奶、盐、温五者相互协调才能煮出美味可口的咸奶茶。蒙古族姑娘从小就学习煮咸奶茶，成年后大都练就了一手烹煮咸奶茶的功夫^①。如有客人拜访，主人会先奉上一碗热气腾腾的咸奶茶，表达对客人的欢迎和尊敬。因蒙古族分布地区较广且多与其他民族杂居，蒙古族咸奶茶制作技艺有多种^②，饮茶习俗亦不尽相同。

Mongolian's day begins with a hearty pot of salty milk tea, which is enjoyed by the family for the whole day. It is usually made from brick tea, fresh milk and salt; some families will make it more delicious by putting

---

① 蒙古族姑娘从小就学习煮咸奶茶，出嫁时，新娘要显露一下煮茶的本领，并将亲手煮好的咸奶茶献给宾客品尝。

② 例如，内蒙古东部区牧民以熬制红茶为主，西部区以砖茶为主；东部区习惯将奶食品与茶混合熬制，西部区则常将奶食品泡在熬制好的奶茶中食用。此外，炒米、黄油、奶油、奶皮子等均可作为佐料；添加的鲜奶也可分为牛奶、马奶、羊奶等。

蒙古族咸奶茶

butter, fried rice, dairy products and even meat. A good pot of milk tea requires the balance of tea, milk, salt and water; besides, the utensils used, timing, and order of adding in are also important. Most Mongolian women are masters of making milk tea since they have been trained when they were little girls. When served to guests, salty milk tea is a symbol of hospitality and respect. Since Mongolians have settled down in different places with other nationalities, the way of preparing salty milk tea differs in different regions.

（4）土家族擂茶 / Mashed Tea of the Tujia Nationality

擂茶，顾名思义就是把茶和一些配料放进擂钵里擂碎后加水做成的茶。擂茶习俗流行于多个省份，不同地区擂茶的做法也有差异[①]。传说擂茶与湘西土家族的起源息息相关。土家族擂茶最基本的原料包括生叶[②]、生姜和生米，故又名"三生汤"。制作时将原料置于擂钵内，捣成糊状，做成擂茶"脚子"，再将"脚子"放入碗内，冲入沸水[③]，

---

① 例如，福建西北部的擂茶是将茶叶和芝麻研成细末，加沸水冲泡而成；广东揭阳、普宁等地的客家擂茶是把茶叶擂成粉末后，加上捣碎的熟花生、芝麻、盐和香菜，用沸水冲泡而成；湖南桃花源一带把茶叶、生姜、生米擂碎，加上芝麻和盐等佐料，冲入沸水后饮用。

② 生叶指新鲜的幼嫩茶芽叶。

③ 水要高冲，且要冲得急，让"脚子"在旋转的水里自然冲匀。

加上炒米，擂茶就做好了。热气腾腾的擂茶色泽黄绿，上下层次分明，吃的时候先喝上层的茶汤，再用调羹舀着粥样的擂茶"脚子"吃。随着生活水平的提高，土家人擂茶的原料更为丰富，最常见的是茶叶、生姜、米、芝麻和花生，简称"茶姜米芝花"，还有原料达十种的"豪华版"擂茶——"十全大补茶"①。

Mashed tea, as the name suggests, is made by mashing all ingredients up in an earthen pot and then making tea with the paste. It is a daily drink in many places, and each has its distinctive recipes. For Tujia people, especially those living in the west of Hunan province, mashed tea is of greater importance since it is related to the origin of the nationality. When Tujia people make mashed tea, they put all ingredients (usually including tea, ginger and rice) into a pot and mash them up. The paste, called "Jiaozi" by the locals, is then moved into a bowl. People will pour boiling water into the bowl and add some fried rice. A bowel of mashed tea is ready! The hot tea is yellow and green in color, with distinct layers from top to bottom. When eating the mashed tea, people first drink tea soup on the top layer, and then eat the porridge-like "Jiaozi" with a spoon .Today, a variety of ingredients are used when making the mashed tea. For example, the common version of mashed tea of Tujia

擂茶（视觉中国　供图）

① 土家族的"十全大补茶"原料一般包括芝麻、核桃、绿豆、猪油、红糖、红枣、黑胡椒、花生、黄豆、生姜等。

people includes tea, ginger, rice, sesame and peanut; there is also a luxury version, which is made of ten kinds of ingredients.

（5）瑶族油茶 / Oil Tea of the Yao Nationality

油茶是瑶族古老的饮食种类，既是瑶族的特色美食，也是待客佳品。油茶的做法与擂茶有相似之处，不同的是做油茶需要用油，或许这也是油茶的名称来源。不同的地区油茶的原料和做法会有差异[①]，其中广西恭城[②]的瑶族油茶最为出名。当地人每天早晨都要打油茶，有的家庭一日三餐都离不开油茶。打油茶的工具独特[③]，工序复杂，包括洗茶、打茶、煮茶、滤茶等[④]，茶叶一般用当地特制的茶叶。

Oil tea is a traditional food of the Yao nationality which is not only a daily necessity of the Yao people but a popular ethnic food serving to guests. There are many similarities between oil tea and mashed tea. The distinctive characteristic of oil tea is the use of oil. There are different recipes for making oil tea in different places. And the oil tea in Gongcheng Yao Autonomous County of Guangxi Zhuang Autonomous Region is the most famous. The locals make oil tea every morning and some families love oil tea so much that they eat it at every meal. The tools of making oil tea

---

[①] 据说古法油茶原料包括茶叶、生姜、蒜、花生，以上原料捶烂之后，加少量鸡油，最后加水。制作的关键在于茶浆的制作、火候和温度的掌握。

[②] 瑶族油茶习俗主要分布在广西桂林市恭城瑶族自治县及龙胜、灌阳、资源和贺州市的钟山、富川等地的瑶族聚居区，湖南及广东的瑶族聚居区也有分布。

[③] 工具包含一个生铁锻造、带有锅嘴儿的油茶锅，一个木制"7"字形油茶槌，一把竹制或者藤制的油茶滤。

[④] 茶叶浸泡后，与生姜等调料倒入锅中，用茶槌反复捶打。捶好后加猪油、水和盐，烧开熬至出味，用竹漏斗把金黄色茶水滤入碗中，再放入炒米、麻蛋果、排散、花生等配料，撒上葱花、香菜等。茶汤被饮完后，可向锅中加水继续熬煮，如此可重复熬煮五六锅。故油茶有"一杯苦、二杯呷、三杯四杯好油茶"之说。

油茶（陈燕　供图）

are full of ethnic characteristics. People there will choose the local tea; after complicated processes such as steeping tea, mashing tea, boiling tea and filtering tea soup, they can enjoy a cup of golden oil tea.

油茶与其说是一道茶，不如说是一碗养生汤。瑶族一直生活在山林地带，潮湿闷热，瑶族人民就地取材，把茶叶、生姜、蒜、油、盐等集于一锅，打出了健康，打出了民族特色，也打出了致富之路。油茶是瑶族适应环境的经验总结，体现了瑶族人民在艰苦环境下乐观积极的生活态度，也是瑶族待客、重要场合和仪式中不可缺少的部分。2021年，"茶俗（瑶族油茶习俗）入选国家级非物质文化遗产代表性项目名录，成为广西文旅的一张名片。

Oil tea is in fact a soup with health properties. The Yao people have been living in mountains and forests since ancient times. The environment is not favorable for living as the climate is humid and muggy. The Yao people are so intelligent that they make good use of local specialties such as tea, ginger, garlic, oil and salt. While making a healthy and delicious daily beverage, Yao people make their life more prosperous. In 2021, the oil tea of Yao nationality, the collective wisdom of Yao people, was included in the national intangible cultural heritage list. Presently, oil tea has become the star attraction of culture and tourism of Guangxi Zhuang Autonomous Region.

（6）德昂族酸茶 / Sour Tea of the De'ang Nationality

德昂族[①]是我国古老的种茶族群之一，曾称"崩龙"族，其民族史诗《达古达楞格莱标》[②]记载："德昂族是茶叶变的，茶是德昂族的根。"有"古老茶农"之称的德昂人将茶树作为图腾来崇拜，视茶树为自己的保护神和始祖神。

The De'ang Nationality is one of the ancient tea growing nationalities in China. Used to be called "Benglong" Nationality, they are believed to be the descendants of tea, as described in their national epic. Known as "ancient tea farmers", they worship tea trees as totems, regarding tea trees as their protector and ancestor.

德昂族在长期的生活劳作中，针对南方湿热的天气，摸索出一套制作酸茶的方法。德昂族选择春夏两季制作酸茶，选用云南大叶种茶树鲜叶为原料，制作工艺流程繁杂[③]。德昂酸茶分为干茶和湿茶，湿茶

---

① 德昂族是云南西南部跨中缅边境而居的少数民族。

② 《达古达楞格莱标》是德昂族民间创世神话史诗，是德昂族以茶为万物之源的思想观念的显现。2008 年，被列入第二批国家级非物质文化遗产名录。

③ 传统的制作工艺流程包括鲜叶拣选（剔除杂质）—清洗鲜叶（除尘）—鲜叶摊凉（沥水）—杀青（蒸青）—揉捻装入竹筒—埋入土坑（2~3 个月）—出坑捣碎—压饼晾晒（1~2 天）—切块曝晒（3~4 天）—成品密封保存等多道工序。

德昂族酸茶

一般当菜食用；干茶则供饮用，其汤色金黄透亮①，滋味酸涩回甘。

In the long search for a beverage which can both satisfy daily needs and is suitable for hot and humid climate, De'ang people finally find the way of making sour tea. Sour tea is usually made in spring and summer with local fresh tea leaves through a complicated process. Sour tea is divided into wet tea and dry tea. The former is generally consumed as a dish; the latter, with pleasant color and unique taste, is for drinking.

---

① 汤色随年份的长短呈现黄绿色、金黄色、红色等不同颜色。

千百年来，酸茶成为德昂族社会交往、防病治病①的重要饮品，承载着德昂人的情感与追求。2021年，"德昂族酸茶制作技艺"被列入国家级非物质文化遗产代表性项目名录。

For thousands of years, sour tea has become a daily necessity of De'ang people. It is indispensable on ceremonial occasions, for cultural recognition, in disease prevention and treatment. In 2021, the processing technique of sour tea of De'ang people was included in the national intangible cultural heritage list.

此外，极具特色的民族茶俗还包括傣族的竹筒香茶、纳西族的盐巴茶与龙虎斗②、布朗族的酸茶、回族的罐罐茶等。这一切都使得我们中华大家庭的茶俗异彩纷呈！

Distinctive ethnic tea customs also include bamboo tube fragrant tea of Dai nationality, salty baked tea and baked tea with liquor of Naxi nationality, sour tea of Bulang nationality, pot-baked tea of Hui nationality, etc. All these ethnic tea customs contribute to the profound and colorful Chinese tea culture!

10.3 中国地方特色茶俗 / Regional Tea Customs in China

中国幅员辽阔、历史悠久、文化灿烂，茶文化亦是丰富多彩。潮

---

① 德昂酸茶富含茶多酚、乳酸菌，不仅让酸茶有了独特的"酸香"风味，还具有清热解暑、消炎解酒、提神润喉、促进消化等作用。

② 龙虎斗指将滚烫的烤茶汤倒入盛放一半白酒的茶杯中，茶酒相遇发出响声，故名龙虎斗。龙虎斗还是纳西族人治疗感冒的传统秘方。

州工夫茶、北京大碗茶、成都盖碗茶、羊城早市茶等地方茶俗，不仅反映了城市的饮茶习俗，也处处透露出城市独特的历史文化和风情。

China is a large country with a long history and splendid culture. As an important part of Chinese culture, tea culture is also brilliant. Distinctive regional tea customs, such as Chaozhou Gongfu tea, Beijing Dawan tea, Chengdu Gaiwan tea and Cantonese Dim Sum Brunch, showcase the history, character and charm of different places.

（1）潮州工夫茶 / Chaozhou Gongfu Tea

潮州工夫茶①指流传于潮汕一带的以乌龙茶为主要用茶，配套精致的泡茶器具，遵照独特程式的一种茶叶冲泡和品饮方式。工夫茶起源于宋朝，在广东的潮州府（今潮汕地区）一带最为盛行，后来在全国各地流行②。

Chaozhou Gongfu tea refers to the traditional tea ritual in Chaozhou and Shantou area, which mainly brews oolong tea with exquisite tea ware and fixed procedures. Gongfu tea originated in the Song Dynasty and prevailed mostly in Chaozhou, Guangdong province (now Chaoshan area). It later became popular around the nation.

---

① 广东省非物质文化遗产"潮州工夫茶"传承人陈香白认为："中国茶道、中国工夫茶、潮州工夫茶是三位一体的。"

② "工夫茶"最早的文字记载见于清代俞蛟著《梦厂杂著·潮嘉风月》："工夫茶，烹治之法，本诸陆羽《茶经》，而器具更为精致。"普遍认为潮州工夫茶始于宋代，也有学者认为，潮州工夫茶的源头是明清时期在闽粤一带兴起的乌龙茶的饮茶风尚。

工夫 [①] 茶指一种需要花费较多时间和程序的精细的泡茶及饮茶方式。工夫茶冲泡技法繁复，对茶叶、茶具、用水、冲法、品饮都有严格要求。

Gongfu tea refers to a sophisticated way of making and drinking tea with given procedures (in Chinese, when people spend much time and energy doing something, we may say people spend much "Gongfu"). It has set standards for every aspect including tea, tea ware, water, brewing skills and the way of drinking tea.

潮州工夫茶的茶具非常讲究，传统工夫茶具有 18 件，其中最具特色的有 4 件，被誉为"茶中四宝"——红泥炉、玉书碨、孟臣罐和若琛瓯（即茶炉、茶铫、茶壶、茶杯）。

The tea set of Chaozhou Gongfu tea is elaborate. There are 18 traditional tea ware, of which the most important four are known as the "four treasures of Gongfu tea", namely the stove made in red-clay, Yushu kettle, Mengchen teapot and Ruochen teacup [②].

潮州工夫茶以乌龙茶为主要茶品，来自潮州的凤凰单丛是当地人最喜欢的茶。完整的冲泡程式包括 20 多个环节 [③]，一般只在正式场合展

---

① 潮州人认为，工夫茶是"工夫"，而不能写成"功夫"，前者包括茶人的素养、茶艺的造诣以及冲泡的空闲，体现为人处世的精细和周到，而后者只是一种本领和造诣。此外，在潮州话中，"工夫"与"功夫"的发音不同，前者念"工（gang）夫"，后者念"功（gong）夫"。

② These names consist of the name of the very craftsman who designed the very tea ware.

③ 完整的冲泡程式包括茶具讲示、茶师净手、泥炉生火、砂铫掏水、榄炭煮水、开水热罐、再温茶盅、茗倾素纸、壶纳乌龙、甘泉洗茶、提铫高冲、壶盖刮沫、淋盖追热、烫杯滚杯、低洒茶汤、关公巡城、韩信点兵、敬请品味、先闻茶香、和气细啜、三嗅杯底、瑞气圆融等 20 多个环节。

潮州工夫茶（视觉中国　供图）

示。而在日常生活中，潮州人饮茶程序比较简捷，主要包括备器、冲泡、出汤、品饮几个步骤。这种可简可繁的品饮程式使得潮州工夫茶既可成为一种表演艺术形式，也可融入人们的日常生活。

Dancong (single bush) tea from Chaozhou's Fenghuang Mountain, a type of oolong tea, is a local favorite to serve during the Gongfu tea ritual. There are more than 20 steps in the formal ritual of Gongfu tea, so it is mainly demonstrated on important occasions. In daily life, locals prefer to drink tea in a simple way, that is, getting the tea ware prepared, brewing the tea, pouring the tea into cups and drinking the tea. Therefore, Gongfu tea can be enjoyed either formally or leisurely in everyday life. Gongfu is also reflected in the way of drinking tea.

潮州工夫茶除了冲泡要工夫外，品饮也需要工夫。主客端起茶杯，一闻其香，二观其汤，三慢品细酌。

When the tea is ready, before taking a sip of the tea, people will first smell the aroma, observe the tea soup, and taste the tea.

喝工夫茶是潮州人日常生活中最平常不过的事了，闲来无事、聚会宴饮、拜访商谈，无不以一壶茶相伴。潮州人将茶叶提高到与主食大米同等的地位，把茶叶叫做"茶米"，用茶厚（浓）、茶薄（淡）来形容人情之厚薄。

In Chaozhou, tea is an indispensable part of local people's daily live. The locals do not start a conversation or conduct business without offering a cup of tea first. For the locals, tea is as important as rice, their staple food, so tea is also called "tea rice". Accordingly, "strong tea" and "light tea" are used to describe different levels of interpersonal relations.

工夫茶不只是一门泡茶的技能，更彰显了潮州人为人处世的细腻之道。一杯工夫茶，是生活的日常，也是精神的寄托，更是海内外潮州人身份认同和族群记忆的载体。作为中国现代茶道的起源之一，潮州工夫茶饱含"和、敬、精、乐"的工夫茶精神，是茶艺美学的体现，是承载茶文化千百年沉淀的"活化石"。2008 年，"潮州工夫茶艺"被列入国家级非物质文化遗产名录 [①]。

Gongfu tea is more than a set of brewing skills; it is part of the character of Chaozhou. As the daily beverage of the locals, it is the symbol of the local culture. Meanwhile it gathers Chaozhou people together no matter how far away they are going. As one of the origins of Chinese modern tea ceremony,

---

① 在 40 多项与茶相关的国家级非遗项目中，"潮州工夫茶艺"是唯一一项茶艺类项目。2023 年，中国国际茶文化研究会授予潮州"世界工夫茶文化之乡"称号。

Chaozhou Gongfu tea embodies the spirits of "harmony, respect, refinement and happiness". It is regarded as the "living fossil" of traditional Chinese tea culture. In 2008, Chaozhou Gongfu tea art was added to the national intangible cultural heritage list.

随着社会的发展、生活节奏的不断加快，大多数年轻人对工夫茶已不甚了解。即使在潮州，传统的工夫茶技艺也逐渐被简化的冲泡程式取代。一方面，我们理解并接受基于"便捷"和"速度"的简化冲泡程式；另一方面，我们期待传统潮州工夫茶技艺的传承与保护得到重视。

Due to the fast pace of modern life, many young people know little of the traditional culture of Gongfu tea. Even in Chaozhou, the birthplace of Gongfu tea, it has been found that the traditional Gongfu tea ceremony is often replaced by simplified versions. On the one hand, we are happy to see the simplified version of Gongfu tea which give priority to "convenience" and "speed" attract more young people. On the other hand, we hope Chaozhou Gongfu tea culture can be better protected and inherited.

（2）北京大碗茶 / Beijing Dawan Tea

"叫一声杏仁儿豆腐，京味儿真美。我带着那童心，带着思念么再来一口大碗儿茶。世上的饮料有千百种，也许它最廉价，可谁知道，谁知道，谁知道它醇厚的香味儿，直传到天涯……"多年前，一曲《前门情思大碗茶》让北京大碗茶家喻户晓。顾名思义，大碗茶就是用大碗盛装供客人饮用的廉价茶水，是北方人饮茶的一种方式。明代，大碗茶就出现在北京街头，并在清末和民国年间较为盛行。

More than twenty years ago, a song called "Missing Dawan tea in

Qianmen" made Dawan tea known by many people. Dawan tea, meaning big-bowl tea, used to be a popular way of tea drinking in the north of China. As early as the Ming Dynasty, people began to drink Dawan tea in Beijing, and the custom became popular in the late Qing Dynasty and the Republic of China.

卖大碗茶的方式主要有两种：一是坐地摆摊，即在街道、码头、闹市等地方，支一块白布篷，放一张木桌和几把小凳，这就可以吆喝做生意了。还有一种更加简单，挑着扁担的"流动茶摊"，扁担的一头是包上棉套的大茶壶，另一头是盖着布的荆条篮子，布底下是几只大碗，讲究点儿的还预备几个马扎，有人喝茶就摆下马扎，请客人坐下喝茶。无论是坐地摆摊，还是流动茶摊，人们既可以喝大碗茶汤解渴，又能歇脚休息，可谓物美价廉、一举两得。

In the past, Dawan tea was sold either in simple tea stalls or by mobile tea stands. The former, with a white cloth canopy, a wooden table, and a few small stools, was usually set up along streets, at docks or in bustling markets. The mobile tea stall was simpler because everything was carried on shoulders with a pole. At one end of the pole was a big tea pot wrapped in a cotton cover and at the other was a basket covered with cloth. Under the cloth were tea bowls, and sometimes a couple of folding stools. When someone drank tea, the folding stools were placed and guests could drink tea and have a rest. More importantly, it was dirt cheap.

中华人民共和国成立后，大碗茶一度从人们视线中消失，再次出现的时候已是 20 世纪 70 年代末。随着知青返城创业，大茶碗以其成本低、需求大、技术含量低等特点成为知青创业的最佳选择。今天北

京最知名的茶馆——老舍茶馆 [①] 的前身就是"青年茶社"，也就是 20
世纪 70 年代末以卖大碗茶起家的知青创业集体。

After the founding of the People's Republic of China, Dawan tea once
disappeared from the market. In the late 1970s, as educated youth returned
to Beijing, they were eager to make a living in the city. Thanks to the low
cost, mass demand, easy making techniques, Dawan tea helped them to start
their career in that difficult time. The most famous teahouse in Beijing today,
Laoshe Teahouse, grew out of the small tea stall selling Dawan tea.

无论在哪个时期，大碗茶都具有平民性和大众化的特点。这是因
为大碗茶的消费者大多是普通百姓，他们对茶叶质量以及冲泡方式、
口感、环境等并没有过高的要求，目的仅仅是为了解渴。所选茶叶是
北方人喜欢的花茶，并且多数是价格低廉的"高沫"，也就是茉莉花
茶的茶叶碎。在冲泡方式上，大碗茶也非常接地气，多用大壶或大桶，
将泡好的茶水做好保温，以便随买随喝；喝茶时，大碗畅饮，痛快
淋漓。

Dawan tea is known for its low price and easy availability. Since most
customers were ordinary people and thirsty passersby, they did not have
many requirements on brewing skills and quality of tea. Scented tea was
often used to make Dawan tea because of its fragrance. "Gaomo" which
means "broken tea leaves of high-grade jasmine tea" was very popular for
its low price, strong taste and pleasant aroma. After the tea was brewed, it

---

[①] 1979 年"青年茶社"由北京大栅栏街道办事处尹盛喜创立。当年，茶用白色搪瓷碗
盛，碗底有"青年茶舍北京大碗茶"字样，热血沸腾的返城青年在前门大声吆喝"二分一
碗大碗茶"。后来"青年茶社"发展为大碗茶商贸集团公司；1988 年依托大碗茶商贸集
团公司，老舍茶馆成立。今天，老舍茶馆环境典雅，京味儿十足，成为北京知名的文旅
名片之一。老舍茶馆在打造地标性茶馆的同时，也没有忘记"老二分"大碗茶，不时用
大碗茶给人们送上温暖与清凉。

was usually kept warm in a very big barrel.

四十多年前，带着老北京市井温热气息的大碗茶由于贴近生活、贴近需求、贴近百姓，受到人们的欢迎。随着人们生活质量的提高，尤其是价廉物美、口味丰富、携带方便瓶装饮料的出现，以解渴为目的的大碗茶逐渐淡出了街头胡同。今天，北京大碗茶通常以公益形式出现，展现北京地方茶文化风情，带给人们对过往岁月的回忆和品味。

40 years ago, Dawan tea became popular because it catered to the mass market. With the improvement of quality of life, affordable, delicious and convenient bottled beverages can be found everywhere. Dawan tea has to bid farewell to the time. Today, Dawan tea is sometimes provided on special occasions, but it is mainly for public welfare, displaying the past local tea culture and bringing memories and taste of old Beijing.

北京大碗茶（视觉中国　供图）

（3）成都盖碗茶 / Chengdu Gaiwan Tea

盖碗茶是流行于中国西南地区的喝茶习俗，因在成都尤为流行，也叫成都盖碗茶。盖碗茶盛行于清代，初时主要流行于上层社会，后逐渐普及，如今已成为大众饮茶方式。

Gaiwan tea is a tea custom popular in southwestern China. Chengdu, a city steeped in tea, displays the spirit of Gaiwan Tea. The custom first got popular in the upper class in the Qing Dynasty, and today Gaiwan Tea can be found and enjoyed in many cities in China.

盖碗茶的茶具由盖、碗、托三部分组成，从实用角度来看，茶托防烫[①]，茶盖既可保温，又可拨开茶叶、方便饮用。盖碗又称"三才碗"，盖为天、托为地、碗为人，体现了中国传统"天地万物皆为一体"的思想。相传，茶盏托是唐代西川节度使崔宁之女在成都发明的[②]。

Gaiwan, a three-piece tea set, consists of a lid, a bowl and a saucer, representing the heaven, the human and the earth respectively. It was allegedly invented in Chengdu by the daughter of a Sichuan governor during the Tang Dynasty. Functionally, the three-piece Gaiwan well satisfies drinkers: With the saucer, one can easily hold the scalding tea bowl, and the lid can not only keep the tea warm, but also remove the tea leaves when one drinks tea.

盖碗茶的冲泡和饮用都十分讲究，通常包括五道程序：净具、置

[①] 通常，茶托中央会围上一小圆环，如此能更好地固定茶碗。

[②] 考古发现，战国时期，人们拿碗喝茶；到南北朝时，有了茶盏、盏托的组合；到唐朝，茶盏和盏托的组合已经出现。

茶、沏茶、闻香、品饮 ①。

The brewing and drinking of Gaiwan tea follows a certain procedure which includes five steps: cleaning the tea ware, putting the tea into the bowl, pouring water into the bowl, smelling the fragrance, and tasting the tea.

盖碗茶的冲泡是一项充满美感的艺术绝技。说到冲泡，就不得不提茶博士 ②，也就是堂倌。茶博士有"手不停、脚不住、嘴不闲"一说。技艺高超的茶博士是茶馆的活招牌，通常他们右手握着长嘴的铜茶壶，左手卡住碗托和白瓷碗，左手一扬，碗托便整整齐齐摆好，茶碗一个个稳稳落入碗托之中，提壶冲水，没有一滴溅到碗外。

The brewing of Gaiwan tea is an artistic stunt full of beauty. When it comes to brewing tea in Chengdu teahouses, one has to mention Dr. Tea, the highly skilled waiter of teahouses. As the busiest man in the teahouse, Dr. Tea keeps working all the time. Dr. Tea is often viewed as the walking advertisement of the teahouse. Holding a long mouthed copper teapot in their right hand, and tea ware in left, they can brew tea for several customers in a second without splashing a drop of water outside the bowls.

成都以生活安逸闲适著称，泡茶馆是成都人最具特色的休闲生活方式。据统计，有"休闲之都"之称的成都拥有近万家茶馆，茶馆数量为全国之最。成都的老式茶馆是个地域特色鲜明的公共空间。一

---

① 净具即烫杯；净具后取3~5克的茶叶放入碗中；再将刚刚煮沸的开水缓缓沿边注入茶碗，盖上茶盖；3~5分钟后，端起茶托，掀开茶盖闻香；最后，用茶盖将表面茶沫刮开，品饮茶汤。

② 过去茶馆里的堂倌不仅手不停、脚不住、一张嘴也是不闲着的，跟客人讲新闻、说段子、带言传话，似乎天下之事，无所不知，所以堂倌又有"博士"的雅号。

些茶馆的营业场所延伸至户外，通常会提供花茶、沱茶、绿茶等茶叶，老茶馆大多使用小木桌、竹靠椅，茶客坐躺随意，晒着太阳，谈天说地。一般准备到茶馆里"泡"上半天的茶客，都是喝二花、三花的老顾客，就是喝二级、三级茉莉花茶。这类茶不但价格便宜，还经得起长时间的浸泡。在竹靠椅嘎吱嘎吱的背景声中，成都人一边喝茶一边打麻将，不时开个玩笑、斗个嘴，这便是四川特有的谈天说地方式——"摆龙门阵"。

Chengdu is known for its easy lifestyle, and it is said that half of the locals live in teahouse. Presently, there are nearly 10,000 teahouses in the city. Traditional teahouses in Chengdu are a public space with distinct regional characteristics. Some extend their business premises to the open air. Scented tea, Tuo tea, and green tea are often offered; small wooden tables and bamboo armchairs are provided. Basking in the sun, customers may idle the day away over a cup of tea. Regular customers generally prefer second and third grade jasmine tea. This type of tea is cheap and can be brewed for several times. In such a comfortable atmosphere, the locals play mahjong while drinking and gossiping, which is the unique lifestyle of Chengdu — "setting up the Dragon Gate Formation".

时代的变化、城市的变迁，使得成都最具特色的公共空间——茶馆发生了不小的变化。今天，室外开放的茶馆越来越少；在大多数茶馆中，自助的开水壶逐渐代替了茶博士的掺茶技艺；手提长嘴铜壶的茶博士则更多出现在商业表演场合。即便如此，盖碗茶依然是最具成都地方滋味的那杯茶。

Down through the years, teahouses have seen many changes. There are fewer open air teahouses. In many teahouses, a thermos will be served along with the tea, which means you have to serve yourself instead of being served

成都人民公园鹤鸣茶社（视觉中国　供图）

by Dr. Tea, who is now busy with commercial performances. Teahouses are a means for locals to display their attitude toward life; meanwhile, they open windows through which outsiders may vividly observe the city and its people.

"泡一泡"成都老茶馆，感受这个城市闲适的气息，是最具成都特色的休闲方式。

When in Chengdu, do as the locals do. Have a cup of Gaiwan tea and enjoy the coziness and leisure of the city.

（4）羊城早市茶 / Cantonese Dim Sum Brunch

中国很多城市都有早茶，其中历史最久、知名度最高的就是羊城早茶。羊城是广州的别称；早茶，顾名思义就是早点配上茶水。

The custom of dim sum brunch can be found in many cities, and among them, the one in Guangzhou enjoys the highest popularity. Yangcheng, literally the city of goats, is the nickname of Guangzhou. The so-called dim sum brunch refers to a hearty breakfast consisting of tea and a variety of snacks or dim sum.

中国饮食文化发达，而以广州为中心的粤菜将山珍海味的鲜美发挥得淋漓尽致，因而赢得"食在广州"的美誉。若想体会"食在广州"的精髓，茶楼是必去的地方。对于广州人而言，茶楼是个多功能的公共空间，既可大饱口福，也可商谈事务、休闲聚会甚至消磨时间。广州茶馆有"三茶两饭"之说，即在一天之内，茶楼提供早茶、午茶、晚茶，以及午饭、晚饭各一次。如此，四方来客可全天在茶楼休闲消费。

China is a gourmet paradise while Guangzhou, with Cantonese cuisine featuring fresh and precious ingredients, is one of the centers of this paradise. If you want to get the essence of Cantonese cuisine, the teahouse is the first place you should go to. For locals, the teahouse is a multifunctional public space where people may have a big feast, settle a business affair, promote friendship, enjoy leisure time or just idle days away. People may stay in the teahouse for a whole day since the teahouse provides "three teas and two meals". Specifically, "three teas" refers to the dim sum brunch, afternoon tea and evening tea while "two meals" includes lunch and supper.

羊城早茶虽然冠了"茶"字，主角却是点心。但不管食客是否喝茶，茶位费是必定要收的。一般茶楼提供的茶包括普洱、铁观音、菊花或菊普等几种。食客也可自带茶叶，但依然需要支付茶位费。

Although literally it is called "breakfast tea", tea is not that important and what attracts people is the variety of dim sum. And that is why it is translated as dim sum brunch. All customers in teahouses are bound to be charged for tea, regardless of whether they order tea or not. Even those who bring their own tea to the teahouse are no exception. The tea menu usually includes Pu'er tea, Tieguanyin tea, chrysanthemum, or Pu'er tea blended with chrysanthemum.

茶不起眼，点心却精彩得很！爱吃、会吃的广州人结合中西饮食特点，将精华呈现于茶楼中。虾饺、叉烧包、干蒸、蛋挞①、蒸排骨、凤爪、马拉糕、糯米鸡等各式点心通常盛放在精美小巧的器皿中。虽然分量不多，却胜在花色丰富。早茶消费丰俭由人，既有点上满满一桌，大快朵颐的；也有选择传统的"一盅两件"，即一壶茶配上两件点心。

As the name suggests, the highlight of the dim sum brunch is the wide variety of tasteful dim sum. Shrimp dumplings, chashao bao (Cha siu bao), dry parboiling, egg tarts, steamed ribs, steamed chicken feet, Mala cake, glutinous rice chicken are some of the most popular dim sum. Usually only a small portion of dim sum is put on an exquisite and small plate, which makes it possible for customers to taste different dim sum. Some will order a lot of different dim sum. There are also budget friendly choices, such as the traditional set menu of "a tea with two plates of dim sum".

---

① 这四样点心号称早茶界的"四大天王"。这一句中提及的点心，因种类多，概括性介绍一下即可，无须一一翻译。

羊城早茶

洗碗是早茶的规定动作。碗中倒热水，先烫洗筷子和勺子，之后用手指滚动茶杯烫洗，最后将洗碗的茶倒入水盅。据说这一套洗碗流程是辨别当地人和外地人的依据之一。烫洗完毕后，食客们开始一边享用点心一边饮茶，他们或拿出报纸，或点开手机，或谈天说地，一顿早茶吃上两三个小时。

Before enjoying delicious dim sum, the locals will first wash all the tableware with hot water. It is believed to be a criterion for distinguishing locals from tourists. After that, customers may enjoy the hearty breakfast. In such a relaxing atmosphere, many people will stay in the teahouse for a couple of hours, eating, chatting, reading newspapers or playing on mobile phones and that is why it is called brunch.

广州人习惯将吃早茶说成"叹早茶"。"叹"字在广东话里是"享受""品味"的意思。广州人"叹"的不仅是点心的丰盛和美味，更是广州人独有的那种闲逸情趣。而早茶所承载的意义也不只是填饱肚子，它是当地特有的社交方式，也是地方饮食文化的一面镜子，是广州这座城市特征的体现。

In Guangzhou, people are "savoring" dim sum brunch instead of "having" it. The word "savor" in Cantonese means "enjoy" or "taste appreciatively". For the locals, they are not only savoring delicious dim sum, but also the unique leisure atmosphere. In other words, the Cantonese dim sum brunch is far more than a meal; it is also a way of social life, a mirror of the local culinary culture, and a reflection of the charm of the city.

"叹早茶"是羊城不容错过的文化体验与享受，有助于人们了解既懂得赚钱又懂得享受生活的广州人；同时也体现了广州开放、务实、进取的城市特征。

"Savoring" Cantonese dim sum is a must for tourists. It helps people better understand the local people who know how to make life richer and easier, and it also reflects the spirit of the city—being open, practical and progressive.

10.4 中国香港和台湾的茶俗 / Tea Customs in Hong Kong and Taiwan, China

（1）香港茶俗 / Tea Customs of Hong Kong, China

中国香港以华人为主要居民，历史上曾被英国统治百余年，故东西方文化在此和谐共存。香港是茶叶的重要消费地，是英国茶文化和岭南茶文化的交汇点。在此基础上，香港结合地方特色，不断创新地方茶文化。

In Hong Kong, Chinese make up the majority of the population. Because of its long colonial past, Hong Kong has been remaining a vital point of contact, where cultures truly meet. Being the intersection of British tea culture and Lingnan tea culture[①], the city is an important market for tea consumption. While influenced by different tea cultures, Hong Kong has built its own unique tea culture.

茶是香港人生活的重要组成部分。香港茶文化的一个重要来源是英国的红茶文化。100多年前，英国人将喝红茶的习惯带到了香港。香港人从原料、茶具和冲泡方式等多个方面入手，开发了各种红茶的新饮法，比如丝袜奶茶、鸳鸯奶茶等港式风味的奶茶，以更丰富的口味和形式进一步促进了红茶的消费。

Tea is the customary beverage for all occasions in Hong Kong. More than 100 years ago, British way of tea drinking was introduced to Hong Kong. In the following decades, the local people who were born with a spirit of innovation developed their own way of tea drinking. Stocking milk tea and Yuanyang milk tea (or coffee milk tea) are two representatives of Hong Kong-style tea, which attract customers with rich flavors and delicious tastes.

百年前，香港的码头附近有很多茶铺，主要面向在码头做苦力的

---

① Lingnan culture refers to the cultures of Guangdong and the neighboring Guangxi and Hainan provinces.

低收入群体，英式奶茶对他们来说，滋味太淡、价格太贵。这个群体需要价格便宜、滋味浓厚的提神饮料，这也是为什么港式奶茶用淡奶[①]的原因，一是淡奶比鲜牛奶便宜，二是奶味更浓厚，可以平衡浓烈的茶汤滋味。此外，高含糖量使奶茶的味道更有满足感的同时还能迅速补充能量。

Back to the first half of the 20th century, there were a lot of teahouses lining along docks which catered for low-income coolies, for whom the British milk tea was too weak and expensive. Since a strong and cheap beverage was needed, the evaporated milk was used, which was cheaper than fresh milk and served as a better counterweight to the strong flavor of the black tea. What's more, the tasteful sugary Hong Kong-style milk tea could help people get a burst of energy.

2017 年，"港式奶茶制作技艺"入选首批香港非物质文化遗产代表作名录。港式奶茶[②]，也叫"金茶"，是一种滋味浓烈的深褐色饮料，选用不同的红茶拼配，经过反复过滤以增加强度，加入牛奶以增加厚度和奶味。

In 2017, Hong Kong-style milk tea was listed in the first batch of "representative works of Hong Kong intangible cultural heritage". Called "Kam Cha", Hong Kong-style milk tea is a tangy and creamy beverage made from different blends of black tea, which are pulled for several times for strength, and then mixed with milk for thickness and creaminess.

---

① 淡奶是将牛奶蒸馏除去一些水分后的产品，有时也用奶粉和水以一定比例混合后代替，英文为 evaporated milk。淡奶乳糖含量较高，奶香味也较浓。炼乳是在牛奶中加入 40%~45% 的糖，再经加热蒸发掉约 60% 的水，英文为 condensed milk 或 sweetened condensed milk，两者都非常适合制作奶茶。

② 2023 年 7 月，中国香港邮政以"港式奶茶制作技艺"为题，发行特别邮票，宣传这项香港非物质文化遗产。

港式奶茶的代表首推丝袜奶茶。丝袜奶茶这一令人遐想的名称，源自过滤茶渣的白色棉纱网，经茶汤浸泡后，颜色与形状犹如丝袜。丝袜奶茶的茶叶选择[①]和冲泡流程颇有讲究。其冲泡流程大致包括调配茶叶、煲茶和焗茶、撞茶、回温及撞奶[②]等五个步骤。不同的店家会在原料、冲泡手法、茶奶比例等方面有各自的特色[③]，但总体而言，丝袜奶茶茶味浓郁、奶香悠久，口感丝滑细密。

丝袜奶茶

Stocking milk tea[④] is the representative of Hong Kong-style milk tea. The name of stocking milk tea comes from the thin cotton strainer that is used to filter tea. After soaking in tea soup for a period of time, it looks like a woman's stocking. Preparing stocking milk tea properly is not easy. The good taste first comes from the careful selection of teas. Preparing stocking milk tea includes five steps: blending teas, boiling and infusing the tea, pulling the tea, stewing the tea again, and adding milk into the tea. Each store has its own recipe, but generally, a good cup of stocking milk tea should keep a balance between fragrance, smoothness, and mellowness.

---

① 比较讲究的店家会选用粗、中、幼茶叶不同比例配搭，以冲泡出最佳口感的茶汤。

② 撞奶指在茶中加入淡奶。

③ 不少奶茶店会选择斯里兰卡的红茶及荷兰的黑白淡奶，并认为这是丝袜奶茶最正宗的原料。但香港现存最古老的港式奶茶铺，创始于 1952 年的兰芳园，则选择澳大利亚奶源和印度阿萨姆红茶作为原料。故，各店有各店的特色。

④ 为让英语读者更好地理解丝袜奶茶，该词的翻译需要做一些补充说明。如：To make a good cup of milk tea, the tea should be carefully pulled for many times through thin cotton strainer that resembles a woman's stocking.

相较于丝袜奶茶，混合了茶和咖啡的鸳鸯奶茶更具香港特色。鸳鸯奶茶集合了咖啡的香味和奶茶的浓滑，既体现了香港中西文化的融合，又展示了香港人的创新个性。

Compared with stocking milk tea, Yuanyang milk tea, another representative of Hong Kong-style milk tea, has more local flavor. It is made by mixing coffee with milk tea. Combining the fragrance of coffee and the smoothness of milk tea, Yuanyang milk tea not only reflects the integration of Chinese and Western cultures in Hong Kong, but also shows the innovative spirit of the city.

港式奶茶是许多香港人日常生活的组成部分，既是港式早茶、茶餐厅晚餐的标配，也是很多人下午茶的首选。为推广香港独有的奶茶文化，香港咖啡红茶协会①创办了围绕着港式奶茶的国际性赛事，即"国际金茶王大赛"。随着赛事影响力的不断提升，港式奶茶不仅是一杯奶茶，更成为香港文化的一张名片。

Hong Kong-style milk tea is a popular local beverage, which can accompany most occasions. In order to promote the unique tea culture of Hong Kong, Association of Coffee and Tea of Hong Kong has established an international competition named "International Kam Cha Competition". With the increasing popularity of the event, Hong Kong-style milk tea has become a local product with global influence.

香港茶文化的另一个来源是岭南茶文化。受广式早茶的影响，香港也有吃早茶的传统。此外，源于岭南地区的凉茶在香港也有一席之

---

① 香港咖啡红茶协会创立于2008年；2009年该协会创办"金茶王大赛"；2010年，"金茶王大赛"升级成为"国际金茶王大赛"，增设海外赛区。

地。凉茶是非茶之茶①，由特定的中草药烹煮而成，具有祛湿降火、解燥消暑和防治感冒的功效。凉茶在香港已有百余年历史，初期凉茶店以家庭作业模式为主。20 世纪五六十年代，凉茶店大多设唱机和电视机以吸引顾客，成为市民消遣娱乐的地方。20 世纪 70 年代凉茶业开始式微，凉茶店转型自救，推出便携式包装凉茶、颗粒冲剂，并大量生产和推广，衍生出新式的凉茶产业。凉茶至今仍是香港市民经常饮用的饮品。2006 年，传统健康饮品凉茶被列入第一批国家级非遗名录。

Tea culture in Lingnan region also contributes to that of Hong Kong. For example, the local people share the custom of breakfast tea (dim sum brunch) as the neighboring Cantonese. In addition, Chinese herbal tea or cold tea from Lingnan region is also a popular beverage in Hong Kong. As a matter of fact, Chinese herbal tea contains no tea but various herbs. It is known for the virtues of removing dampness, reducing inner fire, beating heat and preventing cold. In the early days, Chinese herbal tea shops were mainly family business. In the 1950s and 1960s, they offered merry time to customers with record players and TV sets. In the 1970s, with the decline of Chinese herbal tea, many stores had to develop new products such as bottled herbal tea and herbal tea granules. Today, health properties of Chinese herbal tea are recognized by more people in Hong Kong. In 2006, as a time-honored healthy drink, Chinese herbal tea was listed in the first batch of national intangible cultural heritage.

此外，香港对普洱茶的工艺创新、消费和储藏也颇有贡献。20 世纪 70 年代，香港对普洱茶的需求，在一定程度上推动了云南对后发酵

---

① "非茶之茶"的意思是虽以茶为名，但原料中并不含茶叶。传统凉茶既不冰，口感也不好。目前人们常见的罐装凉茶是为迎合大众口味而改变的凉茶饮料。但凉茶既以茶为名，也算是茶文化的组成部分。

工艺的研究；而香港存老茶的做法及香港老茶的扩散（1997年，不少茶馆老板因移民离港，大量出售存储的普洱老茶），引发了华人社区对普洱茶老茶的追捧。从这个意义上说，香港还是普洱茶文化保存、升华的堡垒。从那时起，普洱茶的地位堪比法国葡萄酒。

In addition, Hong Kong has made great contributions to the technological innovation, consumption preference and storage of Pu'er tea. To a certain extent, the preference of fermented Pu'er tea of Hong Kong in the 1970s promoted the research on pile fermentation technology (mainly an effort to hasten the aging) in Yunnan. In 1997, a huge amount of aged Pu'er tea was offered for sale because of many teahouse owners' emigration, which triggered the pursuit and prosperity of aged Pu'er tea in Greater China. Since then, Pu'er tea, being a domestic equivalent to vantage French wine, has enjoyed more popularity. In a sense, Hong Kong is the headquarters for the preservation and sublimation of Pu'er tea culture.

值得一提的是，香港的茶消费总体比较务实，即便是较为珍贵的普洱老茶，价格也相对合理公道；高档消费的茶馆也并不多见。这或许是因为香港寸土寸金、寸时寸金，不仅店铺租金昂贵，有大把时间泡茶馆的消费群体也有限，务实的市民文化造就了务实的消费市场。

It is worth mentioning that tea consumers in Hong Kong are very practical. Compared with the soaring price of aged Pu'er tea in Chinese mainland, it is reasonably priced in Hong Kong; and most teahouses in Hong Kong are affordable for consumers. The consumption habits are a reflection of the local culture which cherishes both time and space.

茶是香港文化的一个缩影，在深受中西多种文化影响的同时，也影响了中国内地及周边国家和地区。

Tea culture of Hong Kong is the epitome of the culture of the city, which is constantly refreshed by its Chinese root and global influences. Meanwhile, it also has influence on that of Chinese mainland and neighboring countries and regions.

（2）台湾茶俗 / Tea Customs of Taiwan, China

台湾自古以来就是中国领土不可分割的一部分，台湾与大陆同根同源。台湾约 70% 的土地为山地和丘陵，且高山终年云雾缭绕，自然地理环境适合茶树生长。

Taiwan has been an inseparable part of China since ancient times. About 70% of the land in Taiwan is mountainous and hilly, and the mountains are shrouded in clouds all the year round. Its golden geographical position, together with a favorable natural environment, is good for the growth of tea.

台湾地区原不产茶，17 世纪中叶，茶随着来自福建、广东两个产茶、饮茶大省的移民来到台湾。19 世纪初，福建武夷山茶树种及制茶工艺传入台湾，台湾的茶产业逐步兴起。从 19 世纪末至今，台湾茶产业经历了三次重要转折，即 "日本殖民统治时期"[①] 的工业化、外销贸易转向内需市场以及多元化发展。台湾茶产业几经波折，凭借独特的发展路径，将台湾打造为享誉世界的优质茶叶产区。今天，台湾主要生产乌龙茶、红茶和绿茶，以乌龙茶为特色。

It was not until the mid-17th century when immigrants from Fujian

---

① 1895 年，清政府与日本侵略者签订《中日马关条约》，被迫割让台湾岛、澎湖列岛等，台湾就此进入 "日本殖民统治时期"（1895—1945）。正是在这一时期，台湾茶开始扬名国际。

and Guangdong, two major tea producing and consuming provinces, came to Taiwan, did Taiwan began to produce tea. The tea industry of Taiwan prospered at the beginning of the 19th century when tea species and tea processing technology of Mount Wuyi, Fujian province were introduced into Taiwan. Since the end of the 19th century, the tea industry in Taiwan has undergone three important transitional stages, namely, the industrialization during the Japanese occupation period, the transition from export orientation to domestic consumption, and diversified development. With many twists and turns, Taiwan became one of major tea producing regions, famous for its unique high-quality tea. Presently, Taiwan mainly produces oolong tea, black tea and green tea, with oolong tea being the most famous.

台湾茶道有两个源头 [1]：一是潮汕的工夫茶体系，二是日本茶道。台湾茶道源于潮汕的工夫茶，以浸泡乌龙茶为主，探寻出一套讲究器具、冲泡手法和饮茶体验的体系。同时，台湾茶道深受日本茶道的影响。台湾茶人从日本茶道中找到了中国的传统审美，他们从日本采购了大量器具，并通过日本茶道的源流，去寻找其中蕴藏的中国茶道内涵。在此二者基础上，结合地方特色，最终成就了台湾茶道体系。

Taiwan's tea ceremony has two origins. The first one is the Gongfu tea of Chaoshan, from which Taiwan residents explored a system that highlighted tea ware, brewing skills and drinking experience. The second origin is the Japanese tea ceremony, which helped establish the aesthetic system of Taiwan tea ceremony. Tea lovers in Taiwan went to Japan to search for Chinese traditional aesthetics from Japanese tea ceremony and they also purchased a large number of utensils. Based on these two profound

---

① 此观点源自《三联生活周刊》刊载的王恺撰写的《台湾茶道的兴起》一文。

origins, the system of Taiwan tea ceremony is finally set up.

如果说台湾茶道对潮汕工夫茶的继承和日本茶道的学习借鉴是其传统的一面，那么台湾茶业的综合推广体系、新式茶饮的创制、茶具的创新、茶食的研制则是其创新性的最好体现。

If the inheritance from Chaoshan Gongfu tea and the influence of Japanese tea ceremony are the traditional aspects of Taiwan tea culture, the comprehensive promotion system of the tea industry, the invention of various tea beverages and tea ware, and the development of tea food are the best embodiment of its innovation and creativity.

20 世纪 70 年代后，因成本上涨及出口竞争加剧，台湾茶面临危机，出口惨淡。台湾通过一系列措施保障促进茶产业发展。在通过制茶比赛[1]、茶艺比赛等途径提升茶叶品质的同时，还通过茶文化馆[2]、茶园观光和亮点茶庄等形式发展休闲茶业。

In the 1970s, when Taiwan tea industry had to turn from foreign market to domestic market because of rising costs and fierce competition worldwide, the provincial government took a series of measures to protect and promote tea industry. On the one hand, they greatly improved the quality of tea through a series of competitions in tea brewing and tea evaluation; on the other hand, they successfully combined tea industry with leisure industry and tourism industry by promoting new-style teahouses, tea garden and plantation sightseeing.

---

[1] 1975 年台湾地区第一届优良茶比赛举办，以生产冻顶乌龙著称的南投县鹿谷乡选送的产品获得第一名，并以每斤 4200 元成交，远高于外销茶。从此，产制色香味俱全的高品质茶成为各大茶山的第一要务。

[2] 1977 年，台湾地区第一家工夫茶馆开幕。

同时，为了顺应不同客群的消费需求，台湾茶业在传承的基础上，不断创新。一方面保持以干茶贩售为主的传统形态，另一方面不断研制多元化茶产品。不久，品类丰富的包装茶饮料，以珍珠奶茶为代表的新式茶饮等逐渐成为消费主流，极大提升了茶叶经济效益。

Meanwhile, in order to attract young consumers, diversified products are developed. In addition to the traditional tea, a variety of new tea products have been developed or created, such as packaged tea drinks with various flavors and new tea beverages represented by Bubble tea. Thanks to these measures and innovations, the tea industry in Taiwan is flourishing.

对于年轻一代的消费者而言，台湾茶的最佳代言人当属珍珠奶茶。创制于 20 世纪 80 年代的珍珠奶茶 [1] 以其筋道的木薯珍珠和无限的客制化选项而闻名，一经推出迅速风靡世界。珍珠奶茶的成功得益于它简便的制作方式和多样的口感。对于茶底，珍珠奶茶没有严格的限制，从阿萨姆红茶到绿茶再到台湾自产的冻顶乌龙，都可以作为原料，而且珍珠也可以替换为其他配料，多样的搭配方式诞生了千变万化的口感。珍珠奶茶迎合了年轻人的消费喜好，成功实现了茶的消费群体年轻化。

For young people, the spokesperson for Taiwan tea is no doubt Bubble tea (pearl milk tea). Originating in the 1980s, the simple and tasteful Bubble tea is known for its chewy tapioca pearls and endless customization options, and has quickly swept the world. As for the tea used to make Bubble tea, there are a variety of choices such as black tea from Assam, green tea or the local oolong tea. Besides, various ingredients can be added to the cup. As

---

① 珍珠奶茶是在泡沫红茶的基础上添加粉圆，因为煮过的粉圆形如珍珠，故名"珍珠奶茶"；而泡沫红茶融合了热茶、糖浆及冰块，调泡出的红茶泡沫绵细，因制作时有手摇雪克杯（shaker）这一关键流程，故也称手摇茶。

tasteful Bubble tea brings a lot of excitement to consumers, more and more people are indulged in this new tea beverage.

作为台湾最具国际知名度的原创美食——珍珠奶茶（Bubble tea），早在2005年就被收入牛津权威词典，2018年波霸奶茶（Boba tea）也被收录。现象造就词汇，词汇反映现象，台湾的珍珠奶茶用独特的方式展示了中华美食的世界传播，实现了中式创新茶饮的国际化。

Because of its great popularity, the term "Bubble tea" was included in the Oxford dictionary in 2005, and in 2018 a new term "Boba tea" was added. Words come from reality and they also reflect reality. Thanks

珍珠奶茶（视觉中国 供图）

to Bubble tea, Chinese tea and its culture are known and enjoyed by more people in the world.

台湾茶道在茶具上也颇具创新意识。从20世纪80年代开始，茶具增加了公道杯和闻香杯①。茶在壶里泡好后，倒入公道杯，再从公道杯分入小茶杯，小茶杯的数量也不拘于三个。如此，便舍弃了传统的"关公巡城"，也打破了工夫茶固有的"三杯法"。从实际使用的角度而言，公道杯既可以降温，又可以更加灵活地添加茶汤。闻香杯，顾名思义，主要用于闻香，形制往往比一般的茶杯要细高一些，闻香杯

---

① 潮州工夫茶中不用这两者。陈香白认为工夫茶讲究趁热饮用，公道杯的使用，使茶汤几乎变成温暾水；而工夫茶品饮原本就讲究三嗅杯底，因此闻香杯是多此一举。

的使用，在品饮流程上突出了台湾乌龙的香气。1990 年左右，壶承的出现改变了传统的湿泡法，让冲泡在视觉上更加清爽。

In terms of tea ware, professionals in Taiwan are also full of innovation. In the 1980s, fair cups and aroma-smelling cups were invented. When tea is brewed, it is first poured into a fair cup, and then poured to cups for guests. Functionally, the fair cup ensures the same taste of each cup; meanwhile, the procedure which takes a minute or two, allows tea to cool down. Aroma-smelling cups are a bit taller and slenderer than ordinary tea cups which highlight the fragrance of oolong tea made in Taiwan. In about 1990, teapot holder was invented, which made the brewing process more elegant.

此外，台湾还非常重视茶食的开发与创新。台湾居民擅长以茶入食，即将茶调制成茶食或茶菜肴等。台湾的茶食品种丰富、风味独特，例如在台湾名点凤梨酥的基础上开发的凤茶酥（在制作凤梨酥的过程中加入阿里山的高山茶粉）。台湾的茶菜肴口味多样，既有绿茶沙拉、冻顶茶酿豆腐等爽口清新的菜肴，也有白毫乌龙茶炖牛肉、红茶熏鸡、茶香排骨等浓郁口味的菜肴。创意无限的茶食，让茶以更具烟火气的形式进入日常生活。

In addition, Taiwan residents also attach great importance to the innovation of tea food. They have invented various tea snacks with local flavors. For example, based on the symbol of the local snack, that is, the Pineapple Cake, they have invented tea-flavored Pineapple Cake by adding tea powder into the cake. Tea dishes in Taiwan are flourishing in recent years with refreshing dishes such as green tea salad, tea-flavored tofu, and strong-flavored dishes such as stewed beef with pekoe oolong tea, smoked chicken with black tea, and tea-flavored ribs. With so many tea snacks and dishes, people in Taiwan can enjoy tea in differently ways more comfortably and pleasantly at any time and place.

作为最受欢迎的日常饮品，茶与台湾民众的生活息息相关。对于台湾民众而言，茶是一种与生活结合的艺术，增添了人们的生活情趣，是台湾文化的一缕清香。

As the most popular beverage, tea is a way of life for the people in Taiwan. For Taiwan residents, tea is an art of life, an enjoyment of living and a microcosm of the local culture.

小贴士

1. 盛行于金华市磐安县玉山一带的庙会（赶茶场）是入选国家级非遗名录的民俗项目，该项目与茶祭祀相关。相传东晋道教仙师许逊在游历玉山时，用茶为当地百姓防疫治病做出过巨大贡献，百姓感其恩德尊之为"茶神"，建庙立像，四时朝拜。宋代又开设茶场，形成以茶叶交易为中心的"春社""秋社"两季庙会。春社是农历正月十五，届时茶农祭拜茶神，祈求茶叶丰收，茶场内还会上演社戏、挂灯笼、迎龙灯等。秋社在农历十月十五举行，人们拎着茶叶和货物来到茶场赶集，形成盛大庙会，其间还有叠罗汉、迎大旗等活动。

2. 多穆壶源于蒙古族、藏族等少数民族地区，主要用于盛装奶茶、酥油茶。清代满族也用多穆壶盛装奶茶，至清乾隆时期，多穆壶实用功能弱化，逐渐演变为陈设器。

3. 中国台湾地区茶业的创新与发展给处于相似境况的内地茶业提供了很好的样本。（The development and innovation of tea industry in Taiwan provides important enlightenments for the future development of tea industry in the mainland which is facing similar challenges and transitions.）

10.5 其他国家的茶俗 / Tea Customs of Other Countries

茶的相关习俗不仅遍布中国，而且通过古代丝绸之路等商道影响了其他国家和地区。目前世界各国引种的茶种、采用的栽培方法、加工工艺、品饮方式以及茶俗茶礼都直接或间接地源自中国。中国不仅是茶的故乡也是茶文化的摇篮。

Tea-related customs are not only found across China, but also influenced the rest of the world through the ancient Silk Road and trade routes. It can be easily noted that tea varieties, cultivating and processing techniques, and tea customs like tea preparing methods of many countries are more or less influenced by the Chinese. Therefore, it is safe to say that China is not only the birthplace of tea but also the cradle of world tea culture.

茶是个奇妙的东西，它到哪里都能入乡随俗。目前世界上约有 60 个国家种茶，有 160 多个国家有饮茶的习俗。各国饮茶的习惯与各自的历史、地理、气候以及养生理念有关。以日本为代表的东亚国家，历史上深受中华文化的影响，饮茶习俗始终留有中华茶饮的印记；英国、俄罗斯等国偏爱红茶，因为红茶性暖，英国多湿，俄罗斯高寒，饮用红茶最为适宜；而印度、肯尼亚等地的红茶饮用习惯，则是受了当年英国殖民的影响；北非、西非等地是燥热的沙漠，人们饮用绿茶，既可清热去火，还可以补充维生素。

Tea is really a magic leaf which is accepted everywhere. Today, it is grown in about 60 countries and consumed by people from more than 160 countries and regions. Each culture has its own tea-related traditions, a combination of history, geography, climate and perception of health care.

Historically, having been deeply influenced by Chinese culture, East Asian countries like Japan and Korea share many similarities in tea drinking customs. Countries such as Britain and Russia prefer black tea. This is because Britain is wet and Russia is extremely cold, so strong and warm black tea is their best choice. The habit of drinking black tea in countries like India and Kenya is partially the result of British colonialism. In North and West Africa, green tea is consumed to cope with extreme heat of the desert and provide vitamin supplements.

茶以丰富多样的形式活跃于世界人民的生活中，茶既可热饮也可冷饮，既可清饮也可调饮。热饮、清饮源于中国，是数百年来中国最常见的饮用方式，在受汉文化影响较深的日本、韩国等东亚国家较盛行。热饮、调饮是在泡茶时添加佐料，调和后饮用，佐料多为调味品、奶类、水果等。冷饮、调饮一般是在泡茶时添加水果和冰块，亦可添加奶类。调饮方式在世界范围内比较普遍，欧洲、大洋洲、北美、中东、南亚、西亚等地多采用此种饮法。

In different countries, tea is enjoyed in different ways, cold or hot, plain or flavored. Traditionally, Chinese tend to drink warm tea without adding anything into the tea. This has been followed by neighboring East Asian countries such as Japan and South Korea, which historically have been greatly influenced by Chinese culture. People in regions like Europe, Oceania, North America, the Middle East, South and West Asia love to add various ingredients into the tea. And the mixed tea can be made either warm or cold, depending on personal preference.

（1）日本 / Japan

日本有着独特的泡茶技艺和文化，称为"茶道"。日本茶道源于中国，在传承的基础上创新和发展，具有浓郁的民族风情和极高的美学意义。茶道将饮茶与宗教、哲学与伦理相结合，除了物质享受，更注重通过茶道精神来培育道德观念及审美水平。

As a crown jewel of Japan, the Japanese Tea Ceremony[1] is a ritual of preparing and serving Japanese powdered green tea. Originating from China, Japanese Tea Ceremony was inherited and refined by generations of Japanese tea masters. They have combined tea preparation and drinking with religion, philosophy, and ethics, making Japanese tea ceremony full of ethnic flavors and Japanese aesthetics. Japanese Tea Ceremony is more than a bowl of tea but a spiritual and aesthetic discipline for refinement of the self.

禅宗是影响日本茶道精神内涵的重要因素。许多茶道大师都是禅宗僧侣，因为禅宗与茶道的相通之处在于对事物的纯化。日本茶道的开山鼻祖村田珠光[2]、中兴之祖武野绍鸥[3]，将禅宗的哲学内涵与茶道所追求的精神境界结合起来，奠定了日本茶道的基础。而有着"日本茶圣"之美誉的千利休则是茶道集大成者。

It is commonly believed that the essence of the Japanese Tea Ceremony

---

① Chado, also known as Chanoyu, the Way of Tea or Teaism, is commonly referred to as the Japanese Tea Ceremony in English. 为方便理解，日本茶道的英文统一采用"the Japanese Tea Ceremony"表述。

② 村田珠光（1422—1502），一休宗纯禅师（即"聪明的一休"）的弟子。他将茶与禅相结合，开启日本茶道的发展方向，提出"谨敬清寂"之茶道精神。百年后，千利休更改一字为"和敬清寂"，足见其在日本茶道之地位。

③ 武野绍鸥（1502—1555）集茶人、禅僧、连歌诗人身份于一身，是日本茶道继往开来的一代宗师，也是村田珠光的徒孙、千利休的师傅。

and Zen are the same. Many tea masters are Zen monks because both Zen and Japanese Tea Ceremony traditions have simplicity as their guiding principle. The Japanese Tea Ceremony was begun by Murata Shuko, developed by Takeno Joo and refined by Sen Rikyu, the historical figure considered to have had the most profound influence on the Japanese Tea Ceremony.

据说，日本茶道的"四规七则"就源于千利休。"四规"指"和、敬、清、寂"①，乃茶道之精髓。"七则"指的是，提前备好茶，提前放好炭，茶室应冬暖夏凉，室内插花保持自然之美，遵守时间，备好雨具，时刻把客人放在心上等。茶道还有"一期一会"的说法，即每一次相会都无法重来，一生只此一次，故宾主须以最大的诚意来对待。

It is believed that Rikyu proposed the spirit of tea ceremony which consists of four qualities, that is, harmony, reverence or respect, purity or cleanliness and tranquility. Other principles he set forward include 7 rules: Make a satisfying bowl of tea; Lay the charcoal so that the water boils efficiently; Provide a sense of coolness in the summer and warmth in the winter; Arrange the flowers as though they were in the field; Be ready ahead of time; Be prepared in case it should rain; And act with utmost consideration toward your guests. Japanese also believe the concept of "once in a lifetime", which means that the time shared by guests and hosts can never be repeated. Therefore, both should treat each other with the utmost sincerity.

千利休去世之后，他的子孙和弟子分别继承了他的茶道，开启了

---

① "和、敬"是指主人与客人之间应具备的精神、态度和辞仪；"清、寂"则是要求茶室和饮茶庭园应保持清静典雅的环境和气氛。

日本 16—17 世纪黑乐茶碗

日本茶道的流派。表千家、里千家、武者小路千家，各自继承了千利休的茶风，称为三千家，一直保持着日本茶道的正统地位。日本茶道①无论有多少流派，但基本精神即"和、敬、清、寂"没有变。

After the death of Sen Rikyu, a variety of tea ceremony lineages were set up by his descendants and disciples; among them, the Urasenke, Omotesenke and Mushanokōjisenke are the three best-known ones. However, the spirit set by Sen Rikyu is still central to today's Japanese Tea Ceremony.

　　日本人日常饮用以绿茶为主。日本人在绿茶的分类上做到了极致。

The Japanese usually consume green tea, which is further classified into several categories.

　　首先是被视为日本茶文化代表的抹茶②。简单来说，这是一种绿茶粉，但这种绿茶粉可不简单。并非所有茶鲜叶都能制作抹茶。其原料须源于覆茶园，即采摘前几周，茶园须做遮阳处理。如此，茶叶的叶绿素和茶氨酸含量高，能够形成抹茶特有的颜色、口感和覆盖香③。如今，抹茶不仅用于传统茶道，也是日常饮品，亦可作调味之用④。因抹

---

① 日本茶道采用家元制。所谓家元制，指以家元为中心统领某个流派的制度，它是保障各类艺能代代世袭传承的重要制度，被视为日本社会组织的一个缩影。

② 中国相关国标对抹茶的定义："采用覆盖栽培的茶树鲜叶经蒸气（或热风）杀青后、干燥制成的叶片为原料，经研磨工艺加工而成的微粉状茶产品"。

③ 抹茶特有的香气称为"覆盖香"，是指茶树经遮阴覆盖后加工制成的抹茶产品所特有的鲜香细腻或有海苔香的特征香气。

④ 品质好的抹茶通常用于茶道；在蛋糕、冰激凌等食品制作中，经常使用抹茶粉。

茶是粉末状，其冲泡方式与袋泡茶或叶茶不同。需先将抹茶粉与热水调和成糊状，加少量热水后，再用茶筅击打，产生沫饽后，方可饮用。

Matcha is a special type of green tea powder, which is best known for its significance in Japanese tea culture. Not all tea leaves can be used to make Matcha. A few weeks before plucking, the tea bush is shaded from the sun. In response, the leaf produces more chlorophyll and theanine, which contributes to Matcha's bright color, bold intensity and the distinctive aroma of shaded tea. Today, Matcha is used both in these traditional tea ceremonies and as an everyday drink or flavoring[①]. As mentioned, Matcha tea is a kind of tea powder, and the preparation of Matcha is different from that of other teas. It is made by adding hot water to the powder and mix well with bamboo whisk to proper consistency. When Matcha is properly whisked, one can see the froth on the top of the tea.

其次，玉露是日本顶级绿茶，其栽培方法与抹茶类似，即要求在采摘前几周，茶园覆盖遮光布。此外，茶树对土壤和空气也有要求，因此玉露价格昂贵。作为顶级绿茶，玉露的冲泡要求也比较高，尤其需掌握好水温与冲泡时间[②]。一般冲泡水温以50℃~60℃为宜，冲泡3分钟，然后慢慢地均匀分茶。玉露的冲泡虽然费事费时，但当品尝到鲜美的茶汤时，人们会感叹一切都是值得的。

Gyokuro (meaning Jade Dew) is considered a luxury in Japan. The cultivation method of gyokuro is very similar to that of Matcha. Several weeks prior to harvesting, the tea leaves are shielded from sunlight. Because of the added difficulty in shading, the production cost is higher.

---

① The best of Matcha is used for tea ceremonies and the rest is used for making snacks such as cake and ice cream.

② 水温太高，茶有苦味；若冲泡时间不够，茶的甜味就出不来。

日式煎茶（云晴　供图）

When brewing gyokuro, keep in mind that the ideal temperature of water is 50°C–60°C. Let brew for 3 minutes, and then serve into each cup little by little so as to have a uniform mix in each cup. It takes up some time but tasting the unique flavor of gyokuro is worth it.

最后，煎茶^①则是指由在阳光下栽培的芽叶制作而成的蒸青绿茶，也是日本目前产量和消费量最大的茶叶品类。煎茶的加工工艺与玉露相似，区别在于栽培方法。相对粗老的鲜叶或是夏秋两季采摘的鲜叶一般做成番茶，番茶价格便宜，滋味浓郁，咖啡因含量低，适合日常品饮。颇受白领欢迎的玄米茶，就是由番茶与炒米混合制作而成。

Sencha (meaning infused tea in Japanese) is the most popular tea in Japan. It is a loose leaf tea, made from tea plantations open to the sun. The

①　煎茶有广义和狭义之分，广义的煎茶是指用热水冲泡茶叶；狭义的煎茶则如文中所述是绿茶的一种。

process to make sencha is very similar to that of gyokuro, and the only difference lies in the cultivation method. For everyday drinking, bancha is a good choice, since it is a great inexpensive tea, strong in flavor but low in caffeine. Bancha is different from other Japanese green teas because it is made from older and larger leaves or leaves picked in summer and autumn. Genmaicha or Brown Rice Green Tea is made by mixing tea with roasted brown rice. Genmaicha attracts many white-collar workers by its blended fragrance of both tea and roasted rice.

此外，除了各种绿茶，日本人也喝来自印度和斯里兰卡的红茶和中国的乌龙茶。

Apart from the above mentioned teas, Japanese also drink black tea from India and Sri Lanka and oolong tea from China.

（2）韩国 / The Republic of Korea

自中国茶传入朝鲜半岛一千多年来，那里的人们在吸收中国茶文化的基础上，结合本民族特点，形成了独特的饮茶形式与内涵。朝鲜半岛的茶文化及相关礼仪被称为"茶礼"，其特色是将中国的禅宗文化、儒家文化、道家文化与其传统礼节融为一体。

Chinese tea was introduced into the Korean peninsula more than a thousand years ago, and since then, tea has been incorporated into the local custom. Greatly influenced by Chinese tea culture, tea ceremony in Korean peninsular, called Tea Etiquette, is a combination of Chinese Zen, Confucianism, Taoism with its traditional etiquette.

朝鲜半岛茶礼的集大成者是草衣禅师（1786—1866），被当地人誉为"茶圣",其代表作是写于 1837 年的《东茶颂》^①,该书确立了朝鲜半岛茶自身的地位,揭示了朝鲜半岛的茶道精神,标志着朝鲜半岛茶礼的正式形成。

The founder of Korean peninsular tea etiquette is Chouiseonsa (1786–1866), a monk of Zen sect of Buddhism, known as Tea Saint by the locals. *Ode to East Tea,* a poem written in 1837 by Chouiseonsa, established the independent status of Korean peninsular tea culture, and was regarded as the symbol of the founding of Korean peninsular Tea Etiquette.

当代韩国受西方文化影响较深,咖啡是主流的饮品。20 世纪 80 年代,韩国的茶文化开始复兴与发展。每年 5 月 25 日为韩国的茶日,一些地方会举行茶文化祝祭,其中以"五行茶礼"影响力最大。"五行茶礼"是向茶圣炎帝神农氏神位献茶奉礼。

Today, influenced by Western culture, coffee is the most popular beverage in South Korea. Since the 1980s, tea industry and its culture have received more attention in the country. May 25th is South Korea's Tea Day. On that day, certain events and celebrations are held and among them, "the Five Elements Tea Etiquette" is no doubt the grandest. It is held to worship the tea ancestor, also known as Emperor Yan or Shennong.

茶在日本和韩国的表现方式差异很大。日本茶道有着完整的规矩和仪式,有自洽的美学系统,有宗教般严肃的氛围;韩国茶礼强调用身心感受茶,规矩和仪式相对宽松。早期,草衣禅师将佛理与禅修的喜悦融入茶中,在茶禅一味中悟道。今天,人们在品茶过程中,沿袭

----

① 《东茶颂》用诗歌的方式记载了茶树的生态、茶史、中国名茶、朝鲜茶的由来、制茶方式和事茶的不易等。此外,该书强调朝鲜半岛的东茶在色、香、味和药性上不输中国茶。

了这一习俗，认为喝茶是开心的事情，戒律越少，心情越愉快，故品茶的基调是轻松而惬意的。

Tea ceremonies in Japan and South Korea are very different. Japanese Tea Ceremony is known for its complete set of rules and rituals, a unique aesthetic system, and a religious and serious atmosphere. The South Korean tea etiquette emphasizes that tea is an art of experience, so it should be enjoyed both physically and mentally. In the early days, Chouiseonsa integrated the joy of Buddhism and meditation into tea drinking, through which he achieved enlightenment. Today, people enjoy tea with the same perception of Chouiseonsa. They often chat causally and leisurely over a cup of tea, believing the less discipline they have, the happier they will be.

历史上，朝鲜半岛深受中国儒家文化的影响。今天，韩国茶礼以"和、敬、俭、真"①为核心，形成了具有韩国特色的茶文化。茶礼大体可分为日常的生活茶礼和各种仪式中的仪式茶礼。

韩国茶礼（静观　供图）

---

① "和"为心地善良，"敬"为敬重礼遇，"俭"为清廉简朴，"真"为真诚待人。

Historically, the Korean peninsula has been profoundly influenced by Chinese Confucianism. Naturally, its tea etiquette has attached great importance to etiquette, revolving around the spirits of "harmony, respect, frugality, and sincerity". Tea customs vary on occasions of different levels of formalities.

在传统茶礼中，"茶"具有广泛的意义，几乎所有食物都可以入茶。日常生活中，韩国人非常喜欢喝大麦茶，这可能与他们的饮食习惯有关。韩国饮食以烧烤为主，辅以火锅、泡菜等。这些食物往往会给肠胃带来较大负担，而大麦茶具有去油解腻、清肠胃、助消化的功能。

In its traditional tea custom, tea is loosely defined as everything that can be brewed, so a wide range of ingredients are used to make tea. Among them, barley tea is the national drink. Due to the fact that South Korean dishes are usually with strong flavor, barley tea which stimulates the digestive systems, is helpful to keep dietary balance.

在茶品选择上，韩国人喜欢产自本国的绿茶，也饮用进口的红茶和中国的普洱茶。人们大多崇尚清饮，采用简便易行的冲泡法。通常先用大壶泡茶，倒入类似公道杯的水盂中，而后再倒入茶杯中饮用。

Generally, South Koreans prefer domestic green tea, imported black tea and Pu'er Tea from China. People usually brew tea with a teapot and they prefer to taste the pure flavor of tea. A big cup called "Shuigu" is often used, which can be regarded as the Korean fair cup.

韩国茶叶包装

　　韩国茶园多分布在南部^①，茶园面积不大、产量不高。韩国主产绿茶，且以生产有机绿茶为主。相对而言，中国绿茶味浓，日本绿茶味甜，而韩国绿茶则清爽干净。这一份干净正是源于韩国人在茶园管理、病虫害防护以及茶叶加工方面的精细把控。

　　Concentrated in the south of the country, tea plantations in South Korea are small in size, so is the production. They mainly produce green tea. Many tea plantations produce high-quality organic tea, which comes from plants that have been grown without the use of any chemicals. Generally, Chinese green tea tastes strong; Japanese sweet and South Korean clean and clear. It is the high standard in each step including the management of tea plantations, pest control and tea processing that contributes to the cleanness and freshness of South Korean tea.

---

① 全罗南道宝城郡是韩国最大的绿茶生产地，被誉为韩国的"绿茶之都"。

正是在轻松随意的氛围中，在一口口清淡干净的茶汤中，韩国人感受到了率性和自然。

In a word, South Koreans' attitude toward tea is casual and relaxing, which brings leisure and enjoyment to people.

（3）印度 / India

印度是仅次于中国的第二大茶叶生产国，同时也是茶叶消费大国，所生产的茶叶约 80% 都是供国内消费。印度主要有三个产茶区，每个地区的茶都以地名命名，如阿萨姆①茶、大吉岭②茶和尼尔吉里茶。由于茶区的风土不同，各地的茶各具风味。阿萨姆邦是印度主要茶叶产区，产量约占印度茶叶产量的一半。

India is the second largest producer of tea in the world and it is also a big consumer of tea with about 80% of the total production being consumed by the domestic market. There are three major tea producing regions, namely, Assam, Darjeeling and Nilgiri. Each has a different flavor, strength and taste due to different soil, weather and latitudes. Assam is the main tea producing region in the country, accounting for approximately half of India's tea production.

印度人大多饮用阿萨姆茶，印度南方则较多饮用尼尔吉里茶，而大吉岭茶则主要用于出口③。阿萨姆茶的汤色呈深红色、茶味浓烈；大

---

① 阿萨姆茶园多种植高大树木，为茶树遮阳挡风。（Tea estates in Assam grow many other trees to protect the sensitive tea plants from the sun and strong winds.）

② 大吉岭的茶园绝大多数分布在海拔 1000 多米的地区。海拔越高的茶园，中国种茶树越多；海拔较低的地区，阿萨姆种茶树较多。

③ 大吉岭茶的产量占印度茶叶总产量 1% 左右。

吉岭茶以果味著称；尼尔吉里茶则兼有两者的特点。

Generally, most Indians choose Assam tea, while in South India, people prefer Nilgiri tea. As for Darjeeling tea, it is mainly for export. Assam tea is known for its deep red and strong flavor; Darjeeling tea is famous for its pleasant fruity flavor; Nilgiri tea has both the fruity flavor of Darjeeling and the bold flavor of Assam.

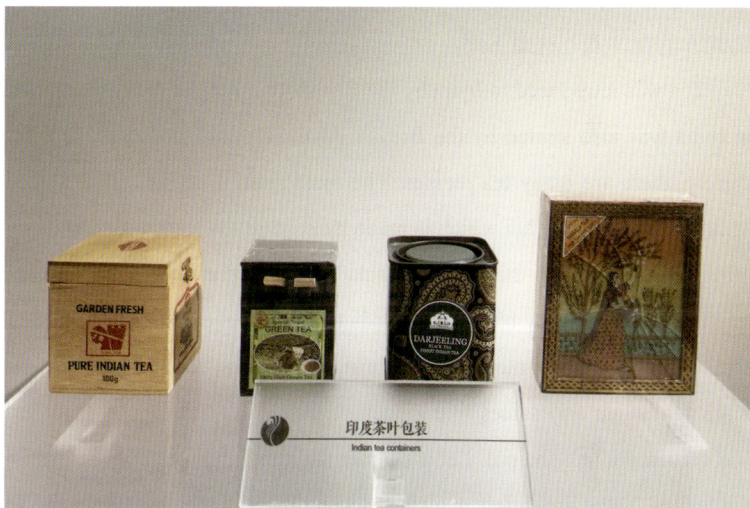

印度茶叶包装

茶是印度人生活的重要组成部分，印度人的一天由茶开启，且他们随时随地喝茶。

India is a true tea drinking community[①]. People there drink tea at any time and place and the day begins with a cup of strong milk tea.

作为曾经的英国殖民地，印度的茶叶种植和品饮受英国影响极

---

① Indians drink eight times more tea than coffee.

大。印度人饮茶以 CTC[1] 红茶为主，且喜欢在茶中添加牛奶。印度的茶饮极具地方特色，其中最特别的当数"香料奶茶"[2]，也称玛莎拉茶（Masla Chai）。在印地语中，"Masala"[3] 是"香料"的意思，"Chai"则指"茶"[4]。制作玛莎拉茶时，一般先煮奶，再加入红茶[5]，再放入糖、生姜片、茴香、丁香、肉桂、槟榔和豆蔻等一起煮几分钟，过滤后即可享用。不同的地方玛莎拉茶的做法各有特点。大致来说，南方讲究"拉"这道工序，也就是用两个杯子将奶茶倒来倒去，以便茶乳交融，而北方重在"煮"这道工序。

Tea was introduced to India by the British, and regular tea consumption in India was also started by the British. Although most Indian tea is CTC variety, there are many tea recipes. The most special one is called Spiced Milk Tea or Masala Chai which is made by mixing tea with milk and many other spices. In Hindi the word "Masala" means spices and "Chai" means tea. To make the milky blend, one needs to follow several steps. First, pour some water and milk into a pot and bring it to a low boil. Then add some black tea, sugar and spices like ginger, fennel, clove, cinnamon, betel nut and cardamom into the pot. Let it simmer for a few minutes and then strain the leaves. There are different tea recipes in different regions. Generally, in the south of India, people will pull Masala Chai with cups to promote the fragrance; while in the north, the tea is often simmered to enhance the taste.

---

① CTC，是"Crush，Tear，Curl"的缩写，属于一种茶叶加工工艺。

② 海底捞火锅推出的"花椒奶茶"算是"擦边"的香料奶茶。

③ 玛莎拉是多种调味料混合而成的复合型调料，可理解为印度的"十三香"，但它的口味更加浓重。

④ "Chai"一词也常常被用来指代玛莎拉茶。

⑤ 也有家庭在制作时，先煮茶，后加奶及各类调料。

印度玛莎拉茶（视觉中国　供图）

　　玛莎拉茶是印度人智慧的结晶。众所周知，很多印度教徒因为宗教的原因只吃素食，因而饮用牛奶是他们补充蛋白质的主要途径；而因为牛在印度文化中的崇高地位，牛奶也被视为吉祥的饮品，被赋予很多美好的寓意。营养丰富的牛奶配上印度盛产的茶、香料和糖[①]，成就了一杯承载印度物产精华的玛莎拉茶。除此之外，玛莎拉茶还是印度人的健康饮品。印度属于热带季风气候，全年高温高湿，而玛莎拉茶里含有多种辛辣的香料。如此，一杯浓郁的玛莎拉茶一方面可以促进食欲，另一方面也可以解热除湿。

　　Masala Chai is a great example of the wisdom of Indians. As we know, many Hindus are vegetarians for religious reasons, and the vegetarian diet may not provide enough nutrition. Since Hindus have long worshipped cows, milk which is an excellent source of protein, is regarded as an auspicious

---

　　① 印度是全球最大的香料生产国，全球70%的香料来自印度；印度还是全球第二大食糖生产国和出口国。

drink, endowed with beautiful connotations. Meanwhile, India is a major producer of tea, spices and sugar. Masala Chai, made by mixing milk, tea, spices and sugar, carries all the essence of India. What's more, it offers many health benefits for Indians. India has a tropical monsoon climate with high temperature and humidity all year round, and spices in Masala Chai can not only arouse the appetite, but also help people cool down by sweating.

玛莎拉茶蕴含着印度各个家庭的味道，各家都有自己喜好和做法，所加香料品种和数量、牛奶和水的比例、茶叶的种类都有差异，这些决定了每家茶的味道也不尽相同。所以玛莎拉茶不仅仅是一款茶，更是一种家的印记。

Every Indian family may have their own recipe to make Masala Chai. The different combination of spices, milk and water creates different strength, flavor and taste of Masala Chai. For some Indians, having a cup of Masala Chai reminds them of the taste of home.

（4）斯里兰卡 / Sri Lanka

自 16 世纪初，斯里兰卡先后被葡萄牙、荷兰和英国入侵。抛开被殖民的屈辱，英国人给斯里兰卡带去的不仅是茶这一最重要的经济作物，还有饮茶的习俗。1948 年斯里兰卡独立前，普通民众较少饮茶；独立后在政府的宣传推动下，饮茶得以普及。

Since the beginning of the 16th century, Sri Lanka was invaded by Portugal, Holland and Britain successively. Despite the humiliating past, the British brought tea, an important cash crop to the island, together with the habit of tea drinking. In the past, tea was not affordable for ordinary people; after the independence of the country in 1948, the government successfully promoted tea consumption.

今天，茶已成为斯里兰卡人日常生活的重要组成部分，各种重要场合都少不了茶。斯里兰卡至今保持着喝下午茶的文化传统，一般是在红茶里加上牛奶和糖。居住在乡间的斯里兰卡人比较喜欢清饮红茶，也会喝生姜茶；城里人除了喝红茶和生姜茶外，还会喝奶茶、水果味冰茶、绿茶、香料茶等。

Today, tea is a daily necessity for the people of Sri Lanka, and it is also an integral part of festivals and important occasions. They have been accustomed to drinking afternoon tea. Generally, they prefer strong black tea mixed with milk and sugar. Those who live in rural areas usually drink black tea or ginger black tea; urban population have more choices such as milk tea, ginger black tea, fruity iced tea, green tea and spiced tea.

虽然 1972 年锡兰国名改为斯里兰卡，但斯里兰卡并没有完全取代"锡兰"一词的使用[1]。150 多年来，锡兰这个名字已经成为世界上最好红茶的代名词。在世人的认知中，锡兰是茶，茶是锡兰。而为了规范斯里兰卡的茶叶出口，该国政府茶叶出口主管机构统一颁发了"锡兰茶质量标志"[2]。这个标志采用了斯里兰卡国旗上的持剑狮王，用以表示斯里兰卡对茶叶质量的承诺。只有经过认可的纯正锡兰红茶，才能使用该持剑狮王标志[3]。

Despite of the change of the country's name in 1972, the tea is still

[1] 该国1948年独立，定国名为锡兰，1972年，改为斯里兰卡共和国；1978年改为斯里兰卡民主社会主义共和国。

[2] 2004年，斯里兰卡茶叶局开始引入"地理标志认证"系统，向国际知识产权局申请了"Ceylon Tea"作为斯里兰卡茶叶的地理标志，2010年获批使用。

[3] 持剑狮王标志诠释了锡兰茶的传统和精髓。有此标志的茶，保证了产品是100%纯正的锡兰茶，在斯里兰卡种植、采摘和包装，符合该国茶叶委员会规定的最高质量标准。

正宗锡兰红茶的狮标

called by the old British colonial name of "Ceylon". For over 150 years, the name Ceylon has become synonymous with the world's finest tea. In the world's eye and tongue, Ceylon is tea and tea is Ceylon. In order to regulate the export of Ceylon Black Tea, the government has officially issued the "Ceylon Tea Quality Mark". The legendary lion of the Sri Lankan flag was introduced to the Ceylon Tea logo to guard this commitment, that is, the symbol of quality. Only the Ceylon Black Tea marked with this logo is the authentic one approved by the Sri Lankan government.

斯里兰卡全年产茶，茶树种植于岛屿的中部高地和南部低地，茶叶按生长的海拔不同分为三类[①]，即高地茶、中地茶和低地茶，全国共有 7 个产区。各产地因风土不同，茶品亦各具特色。高地茶[②]质量最佳，其汤色橙红明亮、口感细腻、回味甘甜、香气浓郁。斯里兰卡各大茶园远离工业污染，坚持不使用任何农药和化肥，以保持茶叶的地道与纯净。

Sri Lanka produces tea all year round. There are 7 tea regions in the central highlands and southern lowlands of the island. Its 7 tea regions, with seven soils and 7 climes, produce a diversity of flavors, aromas, strengths and colors. It is divided into three categories according to the altitude of tea gardens, namely high-grown teas, middle-grown teas and low-grown teas.

[①] 阿萨姆和大吉岭等地的茶叶通常根据季节分类，而锡兰红茶则根据茶园的海拔分类。

[②] 高地茶生长在海拔 1200 米以上的茶园中，中地产区海拔在 600~1200 米，低地产区海拔在 600 米以下。

Among them, high-grown teas[1] are of best quality. With light golden liquor, these teas are fine, smooth and have delicate floral notes. Far away from industrial pollution, the major tea gardens insist on not using any pesticides and fertilizers to keep the tea authentic and pure.

斯里兰卡茶叶包装

斯里兰卡是全球最重要的茶叶生产国之一，主产红茶，也生产少量绿茶和白茶，占全球茶叶产量的 7% 左右。茶叶出口历来是斯里兰卡的主要创汇来源之一。在主要茶叶出口国中[2]，锡兰红茶以优良的品质和口碑，其均价在国际市场中常年保持第一。

---

[1] Teas grow in gardens that are above 1200m in altitude.
[2] 这里指的是年出口量超过 1 万吨的茶叶生产国。

Sri Lanka is one of the world's leading tea producers, accounting for about 7% of global tea production. In addition to black tea, it also produces a small amount of green tea and white tea. Because of its good quality and reputation, Ceylon tea has the highest average price in international market among major tea exporting countries. Tea exports are historically one of Sri Lanka's leading foreign exchange earners.

科伦坡茶叶拍卖市场[①]是世界上规模最大、历史最悠久的茶叶拍卖市场之一，斯里兰卡近 95% 的茶叶都是通过该市场售出。

Almost 95% of the country's tea production is marketed through Colombo Tea Auction, one of the largest and oldest ongoing tea auctions in the world.

（5）马来西亚 / Malaysia

19世纪末，英国人将红茶带入马来西亚[②]；20世纪20年代，福建、广东的华人矿工将茶籽种植在马来西亚土地上；不久印度人将风靡印度的拉茶带到马来西亚。因为马来西亚以伊斯兰教为国教，且伊斯兰教禁酒，茶叶一经推广便迅速普及。

At the end of the 19th century, the British introduced black tea to Malaysia; in the 1920s, Chinese miners from Fujian and Guangdong provinces brought Chinese tea to Malaysia; meanwhile, immigrants from

---

① 斯里兰卡茶叶的产销仓储以其首都科伦坡为中心。科伦坡茶叶拍卖市场创立于1883年，100多年来，基本沿袭当初英国人制定的拍卖制度与流程。

② 马来西亚位于马六甲海峡中心，是海上丝绸之路的重要枢纽和东西方文化交流的据点。早在 600 多年前，郑和下西洋曾五次停留马六甲，中国的饮茶文化最初就是通过马六甲海峡传到马来西亚。

India brought their tea customs to the country. Malaysia is a Muslim country so most people do not drink alcohol according to their religious beliefs. Tea, as a kind of non-alcoholic drink, was welcomed by people once it was introduced to the country.

马来西亚主产红茶，因该地终年高温多雨，茶叶全年可采收。马来人喜好红茶，红茶占国内茶叶消费总量的 80%，其他茶类多为进口[1]。

Thanks to its favorable natural conditions for tea growing, Malaysia produces tea nearly all year round. Black tea accounts for 80% of total domestic tea consumptions. It also imports other kinds of teas.

马来西亚是多种族融合的国家[2]，这也反映在饮茶文化上。受英国人的影响，马来西亚人在喝茶的时候也习惯加奶和糖；当地华人喜欢清饮乌龙茶、普洱茶和绿茶等；马来西亚的饮茶习俗也受到印度的影响。

Malaysia is a multi-cultural country, so its tea culture is diversified. Influenced by the British, most Malaysians drink black tea with milk and sugar; while Chinese Malaysians prefer oolong tea, Pu'er tea and green tea; along with Indian immigrants, Indian tea custom can also be found in the country.

马来西亚人最喜爱的茶饮叫拉茶，也叫飞茶。拉茶[3]源于印度，随

---

① 马来西亚是中国大陆茶叶出口贸易重要的合作伙伴。2021 年，马来西亚是中国第二大茶叶出口国。

② 马来西亚人口中，马来裔占 70.1%，华裔占 22.6%，印度裔占 6.6%，其他种族占 0.7%。

③ 马来西亚拉茶突出"拉"这道工序；而印度的玛莎拉茶则突出其原料中的香料。在制作玛莎拉茶的过程中，部分地区也会采用"拉"这道工序。

着印度人的脚步传至周边国家和地区。目前拉茶是马来西亚、新加坡、泰国等东南亚国家广为流传的特色茶饮。

In Malaysia, the most popular drink is "Teh Tarik", literally "pulled tea". Teh Tarik comes from India. This hot Indian tea beverage has been introduced to neighboring countries and regions and it is very popular in Malaysia, Singapore and Thailand.

拉茶是把泡好的红茶、糖、炼乳倒进器皿里，再用两个器皿来回倒出。为让茶与奶完全融合，拉茶时，器皿间距一米左右，在强烈的撞击中，茶能产生特殊香气且口感绵密。看似简单的拉茶表演既有趣也"危险"，需要表演者高超的技巧和精准度，大开大合犹如舞蹈表演，极具观赏性，因此马来西亚人将其发展为拉茶舞蹈表演。对于许多游客来说，边品茶，边欣赏拉茶表演，别有一番味道。

马来西亚拉茶

Teh Tarik is made by adding condensed milk and sugar into black tea. The locals have the ritual of pouring tea from a great height and this process would be repeated several times to make it frothy and creamy. The artful pour takes practice and skill and it is a delight to watch the tea being pulled so beautifully. Today, making Teh Tarik has been staged for tourists, who are probably attracted both by the performance and the creamy tea.

马来西亚还有一种以茶命名的小吃——肉骨茶，这是一道华人在当地创制的代表性美食①。关于肉骨茶的起源众说纷纭，较为普遍的说法是这道菜起源于海港城市巴生②，据说是巴生的一位中国人于20世纪30年代发明的。肉骨茶用料普通、价格便宜，既可填饱肚子，还有一定的保健功效，因而受到码头苦力的欢迎。

Malaysia is also famous for a snack or dish named Bah Kut Teh, which is believed to be created by the local Chinese. It is said that Bah Kut Teh originated in the 1930s in a harbor city called Klang. In the early days, it first got popular among port coolies because Bah Kut Teh was cheap, delicious and filling. Also it was believed to have health properties.

虽然名为"肉骨茶"，这碗"茶"实则是一道肉骨药膳，不含任何茶叶。之所以称为肉骨茶，一是因为"药"字不吉利，二是因为食用时，人们多会配上一壶茶解腻。配茶通常选福建产的大红袍、铁观音、水仙等乌龙茶类。今天，肉骨茶不再是草根小吃，而是最具马来西亚特色的大众美食。

---

① 马来西亚华人的主体是清末及民国时期自福建、广东、广西及海南等地迁徙而来的移民及其后代。

② 巴生是马来西亚的一个海港城市，巴生港是该国第一大海港。

Sounds strange? How could a tea fill people up? Actually the name of Bah Kut Teh literally translates as "meat bone tea", and at its simplest, it consists of meaty pork ribs simmered in a complex broth of herbs and spices for hours. Despite its name, there is in fact no tea in the dish itself. It is so called for two reasons. Firstly, herbal soup does not sound good since the word "herbal" reminds people of medicine or disease; secondly, Bah Kut Teh is usually served alongside a pot of tea in the belief that it dilutes or dissolves the fat consumed in the dish. People often choose oolong tea from Fujian Province, such as Dahongpao Tea, Tieguanyin Tea, and Shuixian Tea, etc. Today, Bah Kut Teh has been the representative snack of Malaysia.

（6）新加坡 / Singapore

新加坡是一个多民族国家，华人占总人口的 70% 以上，马来西亚人约占 13%，印度人约占 9%[1]。新加坡的饮茶习俗也反映了其民族构成。新加坡不产茶，饮茶的习惯却渗入当地人民的日常生活，其总体饮茶习俗与马来西亚相似。拉茶、肉骨茶也是颇受新加坡民众及游客欢迎的美食[2]。但因华人占比较大，传统的中式茶饮店或中国台湾人开的泡沫红茶店相对较多。此外，由于历史原因，新加坡饮茶习俗也受英国的影响。

Singapore is a melting pot of different nationalities and cultures. The Chinese account for over 70% of the total population; Malays account for 13% while Indians 9%. With the arrival of immigrants from China, India and Malaysia, the cultures of these countries integrated with the local culture.

---

[1]　新加坡华人的祖先较多来自福建、广东和海南省。

[2]　新加坡的肉骨茶与马来西亚的肉骨茶在色香味上各有特点，比如，新加坡肉骨茶一般装在碗中，胡椒味较重；而马来西亚的肉骨茶一般装在砂锅中，药材味较重。

Although Singapore does not produce tea, tea is an integral part of everyday life of Singaporeans. Generally, the tea custom in Singapore is similar to that of Malaysia. Teh Tarik and Bah Kut Teh are also popular in Singapore. However, since most of the populations are Chinese, there are many Chinese-style tea venues in Singapore such as traditional Chinese teahouses and bubble tea shops run by people from Taiwan province. Besides, due to historical reasons, British tea custom also affects the way people consume tea.

（7）泰国 / Thailand

旅游胜地泰国是东盟地区的重要茶叶生产国，主产红茶[1]。泰国大部分地区天气炎热，当地人饮茶以好喝、爽口为目标，喜欢在茶水中加冰块，制成冰茶饮用。泰式冰茶不仅颜值高，味道也很棒，通常采用泰国红茶、牛奶、炼乳为基本原料[2]，香味独特、色泽诱人。泰式冰茶清凉解暑，搭配重口味的泰国菜也是一绝。

Thailand, a world-renowned tourist destination, is a leading tea producer in the ASEAN region and it mainly produces black tea. Most of the country is hot all year round, so the locals love iced beverage. Thai iced tea, or "Cha-yen" is both eye-catching and tasty, which is made from well-brewed black tea, mixed with milk and condensed milk, served over ice. "Cha-yen" is perfect in blisteringly hot weather, and it is a good accompaniment to spicy local food!

---

[1] 在东盟地区，泰国的茶叶产量仅次于越南、印度尼西亚和缅甸。泰国主产红茶，也生产少量绿茶和乌龙茶。

[2] 制作泰式冰茶时，可添加各种香料，包括橙花水、肉桂、八角、茴香和罗望子，也可加入薄荷或各式新鲜水果汁调味。

泰国还有一道与茶有关的特色菜——腌茶。泰国北部与中国云南接壤，饮食习惯与云南少数民族相似，喜酸辣。与云南腌茶一样，泰国人在雨季采摘茶叶，加上盐和香料进行腌制，数月后即可食用。腌茶爽口提神，颇受当地人喜爱。

Besides, there is also a special tea dish, the pickled tea. Thailand borders on Yunnan province in the North, so they share similar dietary customs. For example, they both have an appetite for sour and spicy food. In the north of Thailand, when wet season comes, people will pick fresh tea leaves and preserve them with salt and spices. After a couple of months, pickled tea is ready. Known for its refreshing taste, pickled tea is quite popular with the local people.

（8）英国 / Britain

饮茶是英国人生活的一部分。英国人饮茶以红茶调饮为主，饮茶名目繁多，且对不同时段的茶饮有不同的命名，如起床前喝一杯"床前茶"，清醒提神；早餐时饮用"早餐茶"，搭配培根和煎蛋，去油解腻；工作累了，"茶歇"一下；下午4点，来杯"下午茶"；睡前还有"晚安茶"。在诸多饮茶场景中，最具代表性的当属下午茶。

In Britain, tea is a way of life. They love to mix black tea with sugar and milk. The British take tea seriously, and they even have different names for each cup of tea. For example, "early breakfast tea" wakes them up; "breakfast tea" is a good match for bacon and fried eggs; when they are tired, "tea break" offers a relaxing moment; at 4 o'clock in the afternoon, they relax and enjoy themselves over a cup of "afternoon tea"; and a cup of "evening tea" wishes them a good sleep. Among so many teas, afternoon tea is no doubt the icon of the nation.

英国早餐茶（静观　供图）

据说在 1840 年左右，贝德福德七世公爵夫人安妮·玛丽亚开创了在下午 4 点左右喝"下午茶"的习俗，这一习俗很快流行开来。

It is believed that around 1840, the seventh Duchess of Bedford, Anne Maria, began the popular practice of "afternoon tea", a ceremony taking place at about 4 o'clock in the afternoon. Before long, afternoon tea became quite the social event.

"当钟敲响四下，一切因茶而停止"，下午 4 点到 6 点是正统的英式下午茶时间。茶叶通常选用伯爵红茶、大吉岭茶或锡兰红茶等。喝茶时，人们用茶壶泡好茶倒入杯中，饮用时根据个人喜好加入牛奶和糖。

"When the clock strikes four, everything stops for tea". Traditional afternoon tea is typically served from 4 p.m. to 6 p.m. Black teas such as Earl Gray, teas from Darjeeling or Ceylon are commonly preferred. When brewing tea, people put tea into a teapot, pour boiling water, and then pour the tea into cups. Milk and sugar are added by guests according to their own tastes.

传统英式下午茶

正统的英式下午茶对于茶桌的摆饰、茶具、点心等都非常讲究。茶具精美，通常包括茶壶、茶杯、茶匙、糖罐、奶盅等。英式下午茶确切地说是一顿简餐，最吸引人之处就是丰富的茶点。精美的茶点被放置在三层点心架上。下层放一些咸点心，如三明治、牛角面包等；中层放司康饼、培根卷等传统点心；上层则放蛋糕、水果塔等甜点心。吃茶点的顺序遵循口味由淡而重、由咸而甜的法则，即从最下层的三明治吃起。下午茶讲究的是轻松的氛围，所以用手直接拿点心是流传至今的传统。当然，在日常生活中，下午茶可以不那么讲究，用马克杯冲泡一个茶包，再配上几块饼干就可以了。

Traditional English afternoon tea is very particular about tea table decorations, tea wares, and snacks. Tea wares are exquisite mainly including a teapot, tea cups, tea spoons, a sugar can, a milk cup, etc. It is technically a small meal, with delicious and delicate snacks. Traditional English afternoon tea must be served with three layers of snacks. The bottom tier is the salty sandwich and croissant; middle is the traditional British scones and bacon rolls; top is cake and fruit tarts. The correct way of having snacks should be bottom-up, from salty to sweet. People eat sandwiches first, and then eat up other snacks. Afternoon tea is a time for relaxation, so it is a tradition to take snacks by hand rather than with a knife and fork. Of course, afternoon tea is likely to be quite simple in everyday life, including a couple of biscuits and a mug of tea, which is brewed with a convenient teabag.

英国茶叶包装

今天，嗜茶的英国人随时随地都可以泡上一杯茶，或是消除疲劳，或是沟通交流，或是享受生活。可以说，饮茶习俗影响并在一定程度上塑造了英国人的形象和性格。与炸鱼薯条、板球和英国皇室一样，英式下午茶是英国文化的代名词。

Today, tea is an absolute necessity of British people and they drink tea for various purposes, to relieve fatigue, smooth communications, or delight the soul. To some extent, the image and character of the nation is shaped or at least influenced by tea. Afternoon tea is as synonymous with British culture as fish and chips, cricket and the royal family.

相比于中国，英国饮茶历史不算长，却创造了国际上最知名的茶叶品牌。历史上，英国通过贸易交往和殖民统治影响了诸多国家的饮茶习俗，形成了最具特色和影响力的茶饮文化。

Compared with China, the history of tea drinking in Britain is much shorter. However, Britain has established several world-renowned tea brands. Besides, British tea culture has influenced so many countries that one can never talk about the world of tea without mentioning Britain!

（9）美国 / The United States

作为英国曾经的殖民地，美国早期饮茶习惯深受英国的影响。美国独立之后，茶一度被抵制，之后随着来自世界各国移民的涌入，多种饮茶习俗汇集于美国，形成了独特的饮茶习俗。

Because of its colonial past, the United States was deeply influenced by Britain in terms of tea drinking. For quite a long time after its independence, tea was boycotted. With the coming of immigrants from all over the world, tea customs from different countries were brought to the country.

冰红茶（视觉中国　供图）

美国茶叶几乎全靠进口[1]，其中红茶进口量占总茶叶进口量的 85%。近年来，随着消费者对绿茶健康功效的认可，美国绿茶消费市场稳步扩大[2]；同时，添加水果、香料和花的调味茶也得到消费者的喜爱。

The US does not produce tea, so it has to import a large amount of tea from other countries, of which black

---

[1] 美国是茶叶消费大国、进口大国。近年来，美国几乎从所有的茶叶生产国进口茶叶，红茶主要来自阿根廷和印度；绿茶主要来自中国、阿根廷和日本。

[2] "二战"前，美国人喝的茶 40% 都是绿茶，日本和中国是其主要茶叶进口国。太平洋战争爆发后，美国从日本和中国进口茶叶受阻，便从印度和斯里兰卡进口红茶。从那时起，美国喝红茶的人数开始增加。

tea accounts for about 85%. In recent years, as more and more consumers realize the health benefits of green tea, its consumption keeps increasing; meanwhile, flavored teas or scented teas, teas infused with various fruits, spices and flowers, have also gained increasing popularity.

"速度"和"效率"是美国的文化基因，饮茶也受此影响。美国人普遍不愿多花时间在茶的冲泡上，且人们喝茶如同喝可乐，只关注茶汤口感，极少关注茶叶本身。所以美国人通常用快速便捷和滋味浓郁的茶包泡茶，或者饮用更为方便的罐装茶或瓶装茶。此外，美国人偏好冷食、冰水，喝茶也多喝"冰红茶"，即在红茶中加入柠檬、方糖和冰块后饮用。

"Speed" and "efficiency" are in the gene of American culture, and American tea culture is no exception. People are not willing to spend much time brewing tea, and for most of them, tea is just a beverage which is not different from cola. So most Americans prefer tea bags which are convenient, strong and tasty, or they will choose more convenient canned or bottled tea. In addition, Americans prefer cold food and iced water. Therefore, iced tea is very popular which is made by mixing tea, lemon, sugar and ice cubes together.

据说冰红茶最初是由一位美国茶商于20世纪初发明的[①]。今天，大多数美国人饮用冰红茶，故美国既是冰红茶的故乡，也是冰红茶的天下。

Talking about the iced tea, it is believed that an American businessman

---

① 一般认为冰红茶起源于 1904 年的夏季。一位叫作理查·布利辛登（Richard Blechynden）的茶商在美国圣路易斯市参加世界博览会，因天气炎热，为招揽顾客，该茶商在红茶中加入冰块和糖，冰红茶就此发明并流传开来。

invented iced tea at the beginning of the 20th century. Today, most Americans drink iced tea. Therefore, the US is not only the hometown of iced tea, but also its headquarters.

有趣的是，美国人虽然喝茶不多，却发明了茶包这种最受欢迎的产品形式。1908 年，美国茶商托马斯·苏利文用丝袋将茶叶分装，以便向顾客展示样品。顾客收到茶样后，直接将丝袋置于杯中冲泡，如此便不用费事洗杯子。很快，茶包在世界范围内流行起来，如今美国茶包消费量占茶叶消费总量的 80% 以上。

Interestingly, though Americans are not crazy about tea, they invented the tea bag. It was an American tea seller, Thomas Sullivan, who invented the popular way of brewing tea, that is, using the tea bag. In 1908, Thomas Sullivan sent out free tea in silk bags for people to try. People put the bags together with the tea leaves into the cup and added hot water and they thought it was easy to clean the cup. Tea bags were soon a hit in many countries, and today over 80% of tea consumption is in the form of tea bags.

（10）俄罗斯 / Russia

众所周知，俄罗斯人酷爱伏特加，鲜为人知的是俄罗斯也是一个茶叶消费大国。俄罗斯气候寒冷，虽产茶叶，但产量少，无法满足国内市场需求，因此茶叶主要靠进口^①。

Even though the world is assuming Russians consume bottles and bottles of vodka, it is actually tea which is more frequently drunk all over

---

① 1991年，占苏联茶园种植面积80%以上的格鲁吉亚正式独立，此后俄罗斯茶叶主要依赖进口。近年来，斯里兰卡和印度是俄罗斯最大的茶叶进口国。

the country. It is one of the largest tea consumers in the world. Because of its unfavorable climate, its production cannot meet the domestic needs, it has to import a large amount of tea every year.

俄罗斯有多种茶叶售卖，如红茶、绿茶、花茶等，其中红茶最受欢迎，占茶叶消费总量的 80% 左右。俄罗斯民众对茶叶品牌和口味的要求比较高，如今越来越多的消费者喜欢整叶茶。

There are different kinds of teas available in Russia, such as black tea, green tea and scented tea, of which black tea is the most popular, accounting for about 80% of the total tea consumption. Russians generally prefer big brands and high-quality tea, and in recent years, loose-leaf tea has been favored by more and more customers.

最具俄罗斯特色的饮茶习俗是用"茶炊"① 煮茶。"茶炊"是俄罗斯传统饮茶器具，18 世纪随着茶落户俄国并逐渐盛行。"茶炊"类似煮壶，一般用铜制成，有一个龙头，中间生炭火（类似中国传统的火锅），上面有一个煮茶的茶壶。茶炊的外形多样，有球形、桶形、花瓶形、小酒杯形等，造型丰富的茶炊增添了人们的饮茶情趣。

Russians traditionally use a special kettle called the samovar. As the most representative utensil of traditional Russian tea custom, the samovar was brought into Russia in the 18th century. Usually made of copper, the samovar is a large decorated container for heating water. A central tube running up the middle of the samovar holds charcoal or hot coals necessary for heating the water (the structure is similar to the traditional Chinese hot

---

① 因气候寒冷，为保持茶水的温热，俄国在 18 世纪独创饮茶时的热水器皿——茶炊，饮茶器具随之固定为茶炊、茶壶、茶杯、杯托、茶匙、茶碟、糖块夹等成套的组合。

pot). At the top of the samovar, a crown-like top is made to hold a small tea pot. At the base of the samovar, there is a spigot. People may pour some strong tea from the small teapot then add some hot water by turning on the spigot. With diversified shapes, beautiful and exquisite samovars make tea drinking more enjoyable.

　　饮茶时，人们将茶壶中的浓茶倒入杯中，再打开"茶炊"上的龙头，冲入适量沸水，兑出浓度合适的红茶，再加入糖、柠檬或牛奶。俄罗斯民众喜饮甜茶、浓茶并搭配口味浓郁的茶点①。地处高纬度的俄罗斯冬天天寒地冻，对俄罗斯人来说，喝上一杯浓浓的甜茶是极好的补充热量和取暖的方式。

俄罗斯传统茶饮

When drinking tea, people first pour some strong tea from the teapot and then add some boiling water according to personal preference. They love

---

　　① 通常俄罗斯人选择蛋糕、面包、饼干等黄油、糖分含量高的甜品作为茶点。

to add much sugar, lemons, or milk into the tea and pair it with snacks high in sugar and energy because Russia, which lies in high latitude, is extremely cold in winter. A cup of strong and sweet tea is an excellent solution to the cold weather.

今天，俄罗斯人普遍使用电"茶炊"。传统工艺的"茶炊"逐渐变为装饰品或工艺品，但在重要节日，俄罗斯人会把"茶炊"摆上餐桌，亲朋好友围坐在"茶炊"旁边饮茶，仿佛这样才能体会浓郁的节日氛围[①]。

Today, traditional samovars are replaced by electric kettles. Regarded as something of a ritual, the traditional samovar is still used on special occasions or festivals. On festivals, people will sit with friends and relatives around a warm and reminiscent samovar to celebrate a really good time.

（11）土耳其 / Türkiye

土耳其饮茶之风极盛。2022 年，土耳其以人均饮茶 3.11 千克蝉联人均茶叶消费量最大的国家[②]。土耳其是世界主要茶叶生产国[③]，茶园在其国土的东北部，那里气候与中国南方相似，但冬季气温低，故病虫害少，茶叶在种植过程中不使用任何农药，所产茶叶安全绿色。目前土耳其是世界有机茶种植大国[④]。土耳其生产红茶、乌龙茶和绿茶，红

---

① 近两年，国内秋冬季节流行围炉煮茶（boiling tea on the stove and sitting around it），这种方式与俄罗斯"茶炊"泡茶有异曲同工之处。

② 根据国际茶叶委员会的统计，2022 年，茶叶人均消费量排名全球第一位的仍是土耳其，而中国则以人均茶叶消费量 1.86 千克排名第五。

③ 2023 年，土耳其茶叶总产量居世界第四，仅次于中国、印度和肯尼亚。

④ 恰库（Çaykur）是土耳其最著名的茶叶品牌，统领国内茶叶生产和销售，并且是全球各大茶公司的重要原材料供应商。

茶的生产与消费占有绝对的优势。

Türkiye is well known as a tea drinking country. Statistics show that a single Turk consumes an average of 3.11kgs of tea in 2022, placing them at the top among all other countries. It is one of the largest tea producers in the world. The tea plantations are in the northeast of Türkiye, where the climate is similar to that of southern China, but the temperature in winter is much lower. Therefore, there are few pests and diseases in the tea gardens and no pesticide is used. Most of the plantations produce organic tea. Today, Türkiye is an important producer of organic tea[①]. It produces black tea, oolong tea and green tea and black tea dominates its production and domestic consumption.

土耳其茶叶包装

---

① Çaykur is the most famous tea brand in Türkiye. It is not only the biggest tea producer and seller in the country; it is also a major supplier for many international brands.

茶在土耳其语中为"Cay"发音与中文"茶"极为相似，由此可知，土耳其的茶为经陆路传入的中国茶。早在 16 世纪，茶叶便通过丝绸之路传到了土耳其。20 世纪之前，土耳其人很少喝茶①，20 世纪五六十年代，茶逐渐在土耳其普及。今天，茶被视为土耳其的"国饮"，饮茶成为最日常的生活内容。政府机构、公司或学校通常有专门负责提供茶水的人员；大街上茶馆随处可见②，就连点心店和小吃店也兼卖红茶；红茶"走卖"更是土耳其的特色③。当地有句俗语反映了茶在土耳其的重要性，"交谈时不喝茶如同没有月亮的夜空般索然无味"。对于来土耳其的外国游客，土耳其红茶无处不在，像是一份礼物，如同巴厘岛的花环，也像藏族同胞送出的哈达，有着独特的地方风味，传递着浓浓的情意。

Tea is called "Cay" in Turkish, which obviously stems from the root word "Cha", its Chinese counterpart. In the 16th century, tea was introduced to Türkiye through the Silk Road. Before the 20th century, few people drank tea and it was not until the 1950s and 1960s that tea got popular in Türkiye. Today, as "the national drink", tea has been a way of life for the Turkish. In the offices of government agencies, companies or schools, there are people who are responsible for brewing and serving tea. Outside homes and offices, tea is sold by coffee & tea houses and other small businesses like snacks bars. Besides, there are vendors peddling tea on the street. As a famous Turkish saying goes, "a chat without tea is like a dark sky with no moon". For foreign tourists, wherever they go, tea will be offered as a sign

---

① 20 世纪初，奥斯曼帝国瓦解，原本隶属于土耳其的咖啡种植区脱离了土耳其，土耳其只能进口咖啡，咖啡价格大涨。不久，土耳其试种茶树成功，在政策的提倡下，土耳其人逐渐开始喝本国生产的红茶。

② 受传统和宗教的影响，土耳其女性很少上茶馆喝茶。

③ 在土耳其城市的大街上，穿梭着送卖红茶的伙计，满足人们随时随地喝茶的需求。

of friendship and hospitality, just like a garland of frangipani in Bali and Hada[1] in Xizang.

土耳其红茶品质高、风味稳定，即使长时间泡煮，茶汤也不会变苦。土耳其茶饮最特别之处在于煮茶方式。煮茶使用子母壶，即一大一小两个壶，下方母壶用于烧开水，上方的子壶用来煮茶[2]。斟茶时，将子壶中的浓茶汁倒入玻璃杯中，一般倒四分之一到三分之一杯的量，再将母壶中的沸水冲入杯中，兑出浓度合适的红茶。每个茶杯配一个底盘，上面放着方糖和小勺，可根据自己的喜好加糖，搅拌后趁热饮用。郁金香形的玻璃茶杯也是土耳其茶文化的特色之一，倒入红茶后晶莹剔透。

The local black tea is of good quality. Even if it is brewed for a long time, the tea still tastes soft and mild. In Türkiye, tea comes in a Caydanlik. A Caydanlik has a tea-kettle and a teapot: the kettle for boiling water and the pot for brewing tea. The bigger tea-kettle and the smaller teapot are interestingly compared to mother and child. When the tea is ready, pour the tea into the small tea glasses, usually 1/3 or 1/4 full depending on how dark or how light you prefer your tea. The tea glass is then topped-up with hot water from the tea-kettle. Then the tea glass is placed on a saucer with a small spoon and some sugar. Stir the tea and drink while it is hot. The tulip-shaped tea glass is so eye-catching that it is a symbol of Türkiye tea culture. When the transparent glass is filled with tea, it is crystal clear.

---

① Hada is a piece of silk used as a greeting gift among the Tibetan nationality.

② 用子母壶煮茶时，先冲水至子壶洗茶。在母壶里加满水，将子壶叠放其上。水烧开的过程中，子壶的茶叶通过蒸汽的熏煮慢慢焙出香味。然后把沸水高冲到子壶中，母壶续水，继续小火加热，子壶叠放母壶之上。待水再次煮开后，便可斟茶品饮。

土耳其红茶杯

土耳其子母壶

（12）摩洛哥等西北非国家 / Northwest African Countries Represented by Morocco

西北非国家常年天气干燥炎热，加之人们常年以食牛、羊肉为主，少食蔬菜，而饮茶既能解渴消暑，又能去腻消食，还可以补充维生素类物质，因此这里饮茶习俗盛行。

Influenced by the special geographical location, the climate in northwest Africa is dry and hot all year round, and the diet consists largely of beef and mutton with very few vegetables. Since tea is known for its health properties such as quenching thirst, relieving heat and providing vitamins, it is naturally favored by people in northwest Africa.

以"北非花园"摩洛哥为例，该国信仰伊斯兰教，当地人不饮酒，所以选择茶作为日常饮品。18世纪开始，英国将茶输往摩洛哥，19世纪末，随着茶的大量输入，饮茶在摩洛哥逐渐普及。多年来，茶在摩洛哥人的饮食文化中具有很重要的地位，几乎达到不可一日无茶的地步。

Take Morocco as an example. Known as "Garden of North Africa", Morocco is an Islamic country. As alcohol is outlawed for Muslims, tea is the most popular beverage for the locals. In the 18th century, Britain brought tea to Morocco, which got popular throughout the country at the end of the 19th century. Over the years, tea has become a daily necessity and tea drinking has been a way of life for the locals.

薄荷绿茶是摩洛哥的特色茶饮，原料包括珠茶、绿薄荷和糖。摩洛哥人认为，茶叶使人兴奋，而薄荷有安神之效，两者的配合相得益彰。制作薄荷绿茶①时，先将茶叶和方糖一起放进盛满开水的壶中，煮上3~5分钟，倒进茶壶后，加入几片新鲜薄荷叶；再将茶壶中的茶水从高处倒入杯中，如此既可降低茶汤的温度，也可产生少量泡沫，口感更好。

Hot mint tea is the national beverage of Morocco. The three basic ingredients are Chinese gunpowder green tea②, spearmint and sugar. The locals believe tea is a perfect match for spearmint and since the former makes people excited while the latter helps them cool down. When making mint tea, people first put tea and sugar into a tea-kettle full of boiling water, and then let the tea simmer for 3 to 5 minutes. When the tea is ready, pour the tea into a teapot, and add a few fresh mint leaves. To make the tea frothy and tasteful, people will pour the tea into glasses from a great height, which also helps lower the temperature of the tea.

摩洛哥人喜欢喝浓茶，不仅投茶量大，糖也加得多。当地的糖大

---

① 薄荷绿茶有不同的制作细节。煮茶前一般都会洗茶；煮茶时，有人会将薄荷叶与茶叶、糖同煮；斟茶时，在茶杯里放置新鲜薄荷叶，口感更清爽，视觉效果也更宜人。街头售卖的薄荷绿茶，通常杯子里预先放好薄荷叶、糖块，随卖随做。

② It is so called because the rolled, ball-shaped tea leaves are said to resemble gunpowder pellets.

多是用甜菜提炼的，每块糖 2 千克左右，泡茶前，先用锤子将糖块敲碎备用[①]。摩洛哥人认为，只有当地的糖才能泡出正宗的薄荷绿茶。茶则多选用来自中国的珠茶，唯有诸如珠茶一类的滋味强劲的绿茶，才不会被薄荷夺味。此外，摩洛哥的传统茶具也很有特色，煮茶用银壶[②]，品茶用透明的玻璃杯，使人们在一饱口福的同时一饱眼福。

摩洛哥茶具

---

① 摩洛哥年人均糖的消费量接近 40 千克，远高于世界人均 20 千克的水平。

② 摩洛哥的铜器制造业发达，其茶壶一般是铜制镀银，壶身有伊斯兰风格的纹饰。传统的茶壶和同材质的托盘常在重要场合使用。

Moroccans like their tea sweet and strong. In Morocco, sugar is mainly made from beets. Each lump of sugar is about 2kg and is hammered into small pieces for making mint tea. They believe that the authentic mint tea can't do without the local sugar. And the tea should be strong enough to stand up to the freshness of the mint, so Chinese gunpowder green tea, which is a bold green tea, is often used to make mint tea. In addition, the traditional tea set is also a special feature of the local tea culture. Mint tea is usually made in a silver pot and served in transparent glasses. It is a pleasant experience to see the yellowish green tea topped with fresh green leaves.

薄荷绿茶根植于摩洛哥文化，承载了摩洛哥人热情好客的性格，诠释了他们对生活的理解。通常，薄荷绿茶要连饮三杯，浸泡时间的长短让每一杯茶都有独特的味道，蕴含不一样的意义。第一杯温柔如生命；第二杯热烈如爱情；第三杯苦涩如死亡[①]。

Hot mint tea is the spokesman of Moroccan tea culture, showing the hospitality of the local people. Meanwhile, it conveys Moroccans' understanding of life. It is typically offered for three servings, with each serving having a unique flavor. The first serving is described as being "as gentle as life", the second "as strong as love" and the third "as bitter as death".

摩洛哥饮茶盛行，却几乎不产茶，茶叶依靠进口。摩洛哥是世界上进口绿茶最多的国家，90%以上的茶叶来自中国，多年占据中国茶叶出口排名第一的位置[②]，茶叶已经成为中摩两国的友好使者。

As a large tea consumer, Morocco does not produce tea, so it has to

---

① 讲解薄荷绿茶时，可类比白族"三道茶"。

② 多年来，中国出口的绿茶约有四分之一进入摩洛哥。

import a huge amount of tea every year. For many years, Morocco is the world biggest importer of green tea and over 90% of the imported green tea is from China. And for China, Morocco has been the biggest international customer of Chinese tea. To some extent, tea has witnessed and promoted the friendship of the two countries.

结语 / Conclusion

在这五彩斑斓的世界里，有一样东西可以把所有人联系在一起，这就是茶！每时每刻都有人在饮茶，而在世界的每个角落，都飘着独特的茶香。当茶去到了不同的国家，又催生出了不同的文化，各美其美，而这一切汇聚成我们今天倡导和向往的美美与共。

In a world of cultural differences, nothing can intermingle people together better than tea! Even at this very moment, someone in the world is having a cup of tea. Wherever the place is, there is always a tea custom unique to the country or region. Each retains its own charm while appreciating the beauty of others, which leads to a shared prosperity.

小贴士

1. 我们常听到 high tea 和 low tea，对这两种"茶"的区别，我们可以简单表述为 high tea 不高级，low tea 不低端。The difference between high tea and low tea: Afternoon tea is also known as "low tea", due to the low tables it was served on. While "high tea", also known as "meat tea", was in fact a hot meal of meat pies, vegetables and bread, eaten at the end of the day with a cup of tea by factory workers during the Industrial Revolution. It was called "high tea" due to the high tables it was served on. 所以，简单来说，high tea 和 low tea，主要指的是

桌子的高低。

2. 土耳其人以茶待客时，一般不会介绍茶品，却喜欢在客人面前夸赞自己煮茶的功夫。煮出的色泽红亮的茶叫作 Tavsan Kani，这对水的温度、泡茶时间和水质都有讲究，有兴趣的朋友可参照以下流程尝试。How to make good Turkish tea: For best results, purified water, a tea-kettle and a porcelain teapot are recommended. After rinsing the teapot with hot water, put in one teaspoon of tea per person, while the water is boiled in the tea-kettle. Pour the boiling water from the tea-kettle into the teapot. The flame under the tea-kettle is turned down and the teapot is placed onto the tea-kettle so that it boils with the steam underneath. The tea must brew for 10–15 minutes. The tea is then ready for pouring into the small tea glasses, usually 1/3 or 1/4 full depending on how dark or how light you prefer your tea. The tea glass is then topped-up with hot water from the tea-kettle. The pot of tea should then be drunk within 30 minutes.

非遗中的茶
Tea in Intangible Cultural Heritage

目前，国家级非遗代表性项目名录中与茶相关的项目共48项。其中，与茶相关的制作技艺共42项，大体可分为凉茶（3项）、绿茶（15项）、花茶（4项）、黑茶（10项）、白茶（1项）、红茶（4项）、乌龙茶（3项）、其他（2项）；与茶相关的民俗共6项。详见下表：

序号	项目编号	项目名称	项目类别	公布时间	申报地区或单位
1	413 Ⅷ—63	武夷岩茶（大红袍）制作技艺	传统技艺	2006 年	福建省武夷山市
2	439 Ⅷ—89	凉茶	传统技艺	2006 年	广东省文化厅
3	439 Ⅷ—89	凉茶	传统技艺	2006 年	香港特别行政区民政事务局
4	439 Ⅷ—89	凉茶	传统技艺	2006 年	澳门特别行政区文化局
5	930 Ⅷ—147	花茶制作技艺（张一元茉莉花茶制作技艺）	传统技艺	2008 年	北京张一元茶叶有限责任公司
6	930 Ⅷ—147	花茶制作技艺（吴裕泰茉莉花茶制作技艺）	传统技艺	2011 年	北京市东城区
7	930 Ⅷ—147	花茶制作技艺（福州茉莉花茶窨制工艺）	传统技艺	2014 年	福建省福州市仓山区
8	931 Ⅷ—148	绿茶制作技艺（西湖龙井）	传统技艺	2008 年	浙江省杭州市
9	931 Ⅷ—148	绿茶制作技艺（婺州举岩）	传统技艺	2008 年	浙江省金华市

序号	项目编号	项目名称	项目类别	公布时间	申报地区或单位
10	931 Ⅷ—148	绿茶制作技艺（黄山毛峰）	传统技艺	2008 年	安徽省黄山市徽州区
11	931 Ⅷ—148	绿茶制作技艺（太平猴魁）	传统技艺	2008 年	安徽省黄山市黄山区
12	931 Ⅷ—148	绿茶制作技艺（六安瓜片）	传统技艺	2008 年	安徽省六安市裕安区
13	931 Ⅷ—148	绿茶制作技艺（碧螺春制作技艺）	传统技艺	2011 年	江苏省苏州市吴中区
14	931 Ⅷ—148	绿茶制作技艺（紫笋茶制作技艺）	传统技艺	2011 年	浙江省长兴县
15	931 Ⅷ—148	绿茶制作技艺（安吉白茶制作技艺）	传统技艺	2011 年	浙江省安吉县
16	931 Ⅷ—148	绿茶制作技艺（赣南客家擂茶制作技艺）	传统技艺	2014 年	江西省全南县
17	931 Ⅷ—148	绿茶制作技艺（婺源绿茶制作技艺）	传统技艺	2014 年	江西省婺源县
18	931 Ⅷ—148	绿茶制作技艺（信阳毛尖茶制作技艺）	传统技艺	2014 年	河南省信阳市
19	931 Ⅷ—148	绿茶制作技艺（恩施玉露制作技艺）	传统技艺	2014 年	湖北省恩施市
20	931 Ⅷ—148	绿茶制作技艺（都匀毛尖茶制作技艺）	传统技艺	2014 年	贵州省都匀市
21	931 Ⅷ—148	绿茶制作技艺（雨花茶制作技艺）	传统技艺	2021 年	江苏省南京市
22	931 Ⅷ—148	绿茶制作技艺（蒙山茶传统制作技艺）	传统技艺	2021 年	四川省雅安市
23	932 Ⅷ—149	红茶制作技艺（祁门红茶制作技艺）	传统技艺	2008 年	安徽省祁门县

序号	项目编号	项目名称	项目类别	公布时间	申报地区或单位
24	932 Ⅷ—149	红茶制作技艺（滇红茶制作技艺）	传统技艺	2014 年	云南省凤庆县
25	932 Ⅷ—149	红茶制作技艺（坦洋工夫茶制作技艺）	传统技艺	2021 年	福建省宁德市福安市
26	932 Ⅷ—149	红茶制作技艺（宁红茶制作技艺）	传统技艺	2021 年	江西省九江市修水县
27	933 Ⅷ—150	乌龙茶制作技艺（铁观音制作技艺）	传统技艺	2008 年	福建省安溪县
28	933 Ⅷ—150	乌龙茶制作技艺（漳平水仙茶制作技艺）	传统技艺	2021 年	福建省龙岩市
29	934 Ⅷ—151	普洱茶制作技艺（贡茶制作技艺）	传统技艺	2008 年	云南省宁洱哈尼族彝族自治县
30	934 Ⅷ—151	普洱茶制作技艺（人益茶制作技艺）	传统技艺	2008 年	云南省勐海县
31	935 Ⅷ—152	黑茶制作技艺（千两茶制作技艺）	传统技艺	2008 年	湖南省安化县
32	935 Ⅷ—152	黑茶制作技艺（茯砖茶制作技艺）	传统技艺	2008 年	湖南省益阳市
33	935 Ⅷ—152	黑茶制作技艺（南路边茶制作技艺）	传统技艺	2008 年	四川省雅安市
34	935 Ⅷ—152	黑茶制作技艺（下关沱茶制作技艺）	传统技艺	2011 年	云南省大理白族自治州
35	935 Ⅷ—152	黑茶制作技艺（赵李桥砖茶制作技艺）	传统技艺	2014 年	湖北省赤壁市
36	935 Ⅷ—152	黑茶制作技艺（六堡茶制作技艺）	传统技艺	2014 年	广西壮族自治区苍梧县
37	935 Ⅷ—152	黑茶制作技艺（长盛川青砖茶制作技艺）	传统技艺	2021 年	湖北省宜昌市伍家岗区

序号	项目编号	项目名称	项目类别	公布时间	申报地区或单位
38	935 Ⅷ—152	黑茶制作技艺（咸阳茯茶制作技艺）	传统技艺	2021 年	陕西省咸阳市
39	1183 Ⅷ—203	白茶制作技艺（福鼎白茶制作技艺）	传统技艺	2011 年	福建省福鼎市
40	1513 Ⅷ—267	黄茶制作技艺（君山银针茶制作技艺）	传统技艺	2021 年	湖南省岳阳市君山区
41	944 Ⅷ—161	茶点制作技艺（富春茶点制作技艺）	传统技艺	2008 年	江苏省扬州市
42	1514 Ⅷ—268	德昂族酸茶制作技艺	传统技艺	2021 年	云南省德宏傣族景颇族自治州芒市
43	1014 Ⅹ—107	茶艺（潮州工夫茶艺）	民俗	2008 年	广东省潮州市
44	1014 Ⅹ—107	茶俗（白族三道茶）	民俗	2014 年	云南省大理市
45	1014 Ⅹ—107	茶俗（瑶族油茶习俗）	民俗	2021 年	广西壮族自治区桂林市恭城瑶族自治县
46	1215 Ⅹ—140	径山茶宴	民俗	2011 年	浙江省杭州市余杭区
47	991 Ⅹ—84	庙会（赶茶场）	民俗	2008 年	浙江省磐安县
48	991 Ⅹ—84	庙会（茶园游会）	民俗	2021 年	广东省东莞市

附录 02 茶馆服务英语
English for Teahouse Services

1. 迎宾 / Welcoming guests

（1）欢迎光临湖心亭茶楼！

Welcome to Mid-Lake Pavilion Teahouse!

（2）请问您有预约吗？

Excuse me, do you have a reservation?

（3）我查到您的预约了。您预约的是靠窗两人座，是吗？

I have got your reservation. You wanted a table for two by the window, is that right?

（4）请您跟我来，我为您引座。

Please let me show you to your seat.

（5）地板有点滑，请注意脚下。

The floor is slippery. Please mind your steps.

（6）您想坐哪儿，大厅还是包间？

Where do you want to sit, in the hall or in a private room?

（7）您可以坐在靠窗的这一桌，这里风景别致，是我们茶楼最受欢迎的位置。

You may have this table by the window. It has the best view and is the most popular table among our customers.

（8）包间设有最低消费，您选的这一间最低消费是 4 小时 1200 元。

There is a minimum consumption for private rooms. The minimum consumption for this room is 1200 yuan for 4 hours.

2. 点单 / Taking orders

（1）这是茶单，您看需要点些什么？

This is the tea menu. What would you like to have?

（2）我们茶楼茶品丰富，您可以选择您喜欢的茶品。我也可以为您推荐我们的特色茶品。

Our teahouse offers a wide variety of tea. You may order whatever you like or I can make some recommendations for you.

（3）我们茶馆的特色是普洱茶，您可以试试。

You can try Pu'er tea, which is the specialty of the teahouse.

（4）您可以选择我们的套餐，与单点相比，价格比较实惠。

Our teahouse offers set menu at a more favorable price.

（5）一些老茶客会根据不同的季节和一天当中不同的时间段，选择不同的茶品。

Some experienced guests may choose different kinds of tea according to different seasons and different periods of the day.

（6）品茶时，我们通常会配一些茶点。

We usually prepare some snacks to go along with the tea.

（7）这一款茶点是我们茶楼特制的，搭配普洱最好不过了。

This snack is a specialty of our teahouse and it goes well with Pu'er tea.

（8）先生，您还要点其他什么吗？

Would you like more of anything, sir?

（9）中国茶之所以这么贵，原因很多。我想其中一个原因是，在中国，茶是一种"雅文化"的载体，高档茶叶一般是手工精制的；而在其他大多数产茶国家，茶叶仅仅是饮品，是机器批量生产的商品。

There are many reasons why Chinese tea is so expensive. I think one of them may be that in China, tea is a symbol of elegant life and high-quality tea is often hand-made, which is rare and time-consuming; while in most other tea-producing countries, tea is only a kind of beverage, the product of mass production of machines.

（10）您现在喝的这款普洱茶可以冲泡十次以上，每一泡的滋味都会有些许变化。我们日常所喝的袋泡茶只能冲泡一两次。这样看来，这泡普洱茶是不是就没那么贵了。

The Pu'er tea you are drinking now can be brewed for more than ten times, and the taste will change slightly each time. In contrast, tea bags we often use can only be brewed once or twice. If you think about it this way, the Pu'er tea you are drinking might not seem that expensive.

（11）我们相信随着中国茶及中国文化的传播与推广，会有越来越多的人了解中国茶、爱上中国茶。

We believe that with the spread and popularization of Chinese tea and Chinese culture, more and more people will learn about Chinese tea and fall in love with it.

3. 过程服务 / Serving the tea

（1）虽说泡茶就是把水倒进茶壶，但是要冲泡好一壶茶也不容易，需要冲泡者具备专业知识和冲泡经验。

Though brewing tea is in essence pouring water into the teapot, it takes a lot of professional knowledge and experience to do it well.

（2）不好意思，让您久等了。这是您点的普洱茶和茶点，请您慢用。

Sorry to keep you waiting. This is the Pu'er tea and the snacks that you ordered. Please enjoy.

（3）您是要自己冲泡还是我们工作人员协助冲泡。我们有专业的茶艺师，可以表演茶艺。

Would you like to brew the tea by yourself or do you prefer the assistance of our staff? Our teahouse provides a performance of tea-brewing ceremony by professional staff.

（4）您是喜欢淡茶还是浓茶?

How do you like your tea, weak or strong?

（5）茶已泡好，茶水很烫，请稍等片刻。

The tea is ready but it is very hot. Please wait a moment.

（6）请先闻茶香，再观察茶汤的颜色。

You may first smell the fragrance and then observe the color of the liquor.

（7）品茶时不要急于咽下，可以慢慢感受茶汤在口腔中的滋味。

While drinking the tea, remember to savor it slowly, and not swallow it in one go.

（8）给您添一点开水吧?

Do you want more hot water?

（9）如果您想自己尝试冲泡，请您告诉我，我可以讲解并演示。

If you want to have a try at brewing, please let me know. I will explain the procedures and demonstrate for you.

（10）您自己冲泡时，请务必小心开水（别烫着）。

Please be mindful of the hot water when brewing tea by yourself.

（11）茶已沏好，如果您觉着茶太浓，可以加点水。

The tea has been poured. If it is too strong for you, you may add some water.

（12）您如有任何问题或需要，可以随时叫我，或者您可以按桌上的服务按钮。

If you have any questions or needs, please let me know. Or you may press the call button on the table.

（13）我们的工作就是让您感到舒适。

It is our job to make you feel relaxed and at home.

（14）这是您的账单，您可以选择现金、银行卡、微信或支付宝支付。

This is the bill. You may pay in cash, by credit card, WeChat or Alipay.

（15）这是您的信用卡和发票，请您收好。

This is your credit card and the receipt. Please keep it.

4. 送客 / Seeing guests off

（1）请问您对我们的茶品和服务满意吗？

How did you like our tea and service?

（2）非常高兴您喜爱我们的茶品和环境。

We are very glad to know that you like the tea and the environment here.

（3）您能就如何提供更好的服务给我们一些建议吗？

Can you give us some advice on how to improve our service?

（4）这是我们茶楼的纪念品，希望您喜欢。

This is a souvenir of our teahouse, I hope you like it.

（5）请不要遗忘随身物品。

Please don't forget your personal belongings.

（6）谢谢光临，希望能再次为您服务！

Thank you for coming! We hope to see you again.

参考文献
References

陈宗懋，杨亚军.中国茶经［M］.上海：上海文化出版社，2011.

陈锦娟，Mesut Keskin.跨文化视域下中国与土耳其茶文化比较研究［J］.世界农业，2018（12）.

邓时海，耿建新.普洱茶续［M］.昆明：云南科技出版社，2005.

冈仓天心.茶之书［M］.谷意，译.济南：山东画报出版社，2010.

耿晓辉.茶道与文学［M］.北京：东方出版社，2018.

何有明.茶文化旅游与茶业经济［J］.现代营销（信息版），2019（9）.

李韬.茶里光阴：二十四节气茶［M］.北京：旅游教育出版社，2017.

林语堂.生活的艺术［M］.越裔，译.长沙：湖南文艺出版社，2017.

刘启贵.茶艺师（初级）［M］.北京：中国劳动社会保障出版社，2007.

刘启贵.茶艺师（中级）［M］.北京：中国劳动社会保障出版社，2008.

刘启贵.茶艺师（高级）［M］.北京：中国劳动社会保障出版社，2008.

罗婧，顾颖颖，刘易成，等.不同茶叶中茶多酚和咖啡因成分的对比分析［J］.贵州茶叶，2013（10）.

戎新宇.茶的国度：改变世界进程的中国茶［M］.上海：上海交通大学出版社，2019.

屠幼英，何普明.茶与健康［M］.杭州：浙江大学出版社，2021.

王猛，仪德刚.蒙古族奶茶制作工艺考释及其现代调查研究［J］.农业考古，2016（5）.

王旭烽.茶文化通论——品饮中国［M］.杭州：浙江大学出版社，2020.

文基营.红茶帝国：从产地到品牌，有关红茶的一切［M］.殷潇云，曹慧，译.武汉：华中科技大学出版社，2016.

吴钩.风雅宋：看得见的大宋文明［M］.桂林：广西师范大学出版社，2018.

吴觉农.茶经述评［M］.北京：中国农业出版社，2005.

吴之远，耿晓辉.茶道与文学［M］.北京：东方出版社，2018.

许玉莲.当代茶人努力的方向［J］.茶道，2023（7）.

张正竹，李大祥.茶学专业英语［M］.北京：中国轻工业出版社，2020.

周宝才，丁然，魏新林，等.茶多糖的健康功效研究进展［J］.中国茶叶加工.
  2019（4）.

上海市闵行区茶叶学会.海上茗谭［M］.上海：上海科学普及出版社，2019.

Kakuzo Okakura. The Book of Tea [M]. Putnam's: New York, 1906.

Rong X Y. Tea Nation [M]. Frankfurt: CBT China Book Trading GmbH, 2018.

https://www.pureceylontea.com/.

https://www.chengdu.gov.cn/english/new/2022-12/16/content_e1a50eb297784b5fbaa
  4e332902bf8f0.shtml.

http://www.liptontea.ca/article/detail/176923/tea-flavonoids.

https://www.unilever.com.au/brands/food-and-drink/lipton.html.

https://www.doc88.com/p-441547819223.html.

https://wenku.baidu.com/view/d382a250915f804d2b16c1bc.html.

https://www.teavivre.com/info/chinese-keemun-black-tea.html.

https://www.twinings.co.uk/about-tea/how-is-tea-made.

https://www.bangli.uk/post/5351.

https://www.wikihow.com/Brew-Kung-Fu-Tea.

http://www.puercn.com/cwh/cywh/126457.html.

http://www.puercn.com/cwh/sjcwh/86193.html.

https://www.puercn.com/czs/cybk/129884.html.

http://www.nongyie.com/view-36348-1.html.

http://www.nongyie.com/view-36343-1.html.

https://www.sohu.com/a/378510522_120094506.

http://www.ishuocha.com/lishi/csg/10641.html.

https://zhuanlan.zhihu.com/p/25179272.

http://www.168tea.com/87769-NzE2MTk1MDQtea.html.

http://cn.koreanbuddhism.net/bbs/content.php?co_id=0202.

http://www.nxkg.org.cn/index.php?m=content&c=index&a=show&catid=15&id=662.

https://wenku.baidu.com/view/a89a24eeb84cf7ec4afe04a1b0717fd5360cb285.html.

https://www.sohu.com/a/160495323_99916158.

https://www.sohu.com/a/201551779_100052764.

https://www.puercn.com/chayenews/csfx/196358.html.

http://www.hxytea.com/News/Detail-130.html.

https://www.sohu.com/a/78009606_362854.

https://www.sohu.com/a/195792153_303717.

https://view.inews.qq.com/a/20210118A01U0G00.

https://www.163.com/dy/article/H6BVVHNM0552CPIM.html.

https://www.sohu.com/na/470313735_486911.

https://www.allaboutturkey.com/tea.html.

https://www.163.com/dy/article/GIE9VM5805501EMW.html.

https://z.hangzhou.com.cn/2020/rwwhql/content/content_7731650.htm.

https://www.iqiyi.com/w_19s7vhr7vl.html.

https://www.zhihu.com/zvideo/1352364538931695617.

http://www.puercn.com/cwh/cccd/159384.html.

https://p.51vv.com/vp/sN23dEVa.

https://www.sohu.com/a/418967048_349225.

https://www.hklanfongyuen.com/.

https://www.lcsd.gov.hk/CE/Museum/ICHO/zh_CN/web/icho/the_first_intangible_cultural_heritage_inventory_of_hong_kong.html.

http://www.chinadaily.com.cn/hkedition/2011-04-28/content_12408625.htm.

https://www.zhihu.com/zvideo/1352364538931695617.

http://www.moa.gov.cn/govpublic/ncpzlaq/201807/t20180720_6154464.htm.

https://www.zgbk.com/ecph/words?SiteID=1&ID=207052.

https://baijiahao.baidu.com/s?id=1807164223473061666&wfr=spider&for=pc.

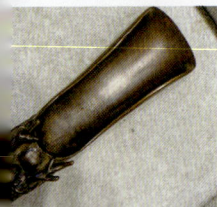

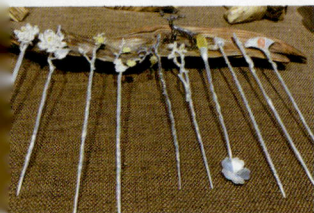